THE ARCHETYPAL KINGDOM

The Dawning of a New Era of Understanding and Creation

THE ARCHETYPAL KINGDOM

by

Ann Ree Colton

ANN REE COLTON FOUNDATION
Post Office Box 2057
Glendale, California 91209

First Edition
ISBN: 0-917189-07-8
Library of Congress Catalog Card Number: 88-70974

For information regarding the writings and teachings of Ann Ree Colton and Jonathan Murro, write:

Ann Ree Colton Foundation
336 West Colorado Street
Post Office Box 2057
Glendale, California 91209
Telephone (818) 244-0113

DEDICATED

TO

ALL WHO TEACH

THE REALITY

OF THE KINGDOM

OF GOD.

. . . behold, the Kingdom of God is within you.
—St. Luke 17:21

CONTENTS

FOREWORD

by Jonathan Murro

I was graced with the opportunity to accompany Ann Ree Colton for over thirty years as her husband and co-worker. Since her passing on June 28, 1984, it has been my joy to place in new books a number of select passages from her voluminous files. THE ARCHETYPAL KINGDOM combines her hitherto unpublished manuscript with her published writings on *the Self-Genesis Age, the Eternal Atoms, the Greater Archetypes, and other important subjects.*

The author's inspired explanations of *the seven Genesis levels* in the first section of THE ARCHETYPAL KINGDOM describe the progressive stages through which man is evolving in this Solar System. The many dynamic truths regarding *Atom Evolvement* will inspire the reader to work directly with God in the quickening of the eternal, spiritual atoms within his being through which comes the Crown of Illumination.

The legacy of Ann Ree Colton includes a vast treasure chest of sacred truths regarding the Transcendental Worlds called in the Holy Bible *the Kingdom of Heaven* and *the Kingdom of God*. The second section of THE ARCHETYPAL KINGDOM is filled with priceless knowledge of the

Beings and Presences dwelling in the Three Heavens; also, there is a thorough description of the little-known *Astral World*. An invaluable Glossary of new words introduced by the author, as well as other words necessary to explain the spiritual mysteries of God, may be found in the final chapter.

Ann Ree Colton's beautiful spirit of wisdom and love lives in her illumined writings and also in the comprehensive System of Niscience, a nonprofit foundation established by her in 1953 for the purpose of blending religion, philosophy, science and the creative arts. Over the years, an increasing number of Truth-seekers have come to recognize the great value of her contributions to the world through her enlightened communion with God and the Lord Jesus. Her love for the Sacred Scriptures of the East and the West is the basis for the many realizations, prophecies and revelations contained in THE ARCHETYPAL KINGDOM.

A new era awaits man, an era in which he will work consciously with the *Archetypes* or *Blueprints* of God for the Creation of the Earth and the Cosmos. As a prophet and revelator under the Christ, Ann Ree Colton has planted many powerful seed-truths that will inspire receptive hearts and fertile minds to understand the spiritual destiny of man as a co-creator with God and an heir of His Kingdom.

> *It is the Father's good pleasure to give you the kingdom.*
>
> —*Saint Luke 12:32*

Part One
Atom Evolvement and Self-Genesis

1

THE NEW-ERA DHARMA

Carry, carry the Dharma with honor. Carry the Dharma with love. Carry the Dharma with courage.

The Dharma wisdom is an eternal thread of deathless spiritual enlightenment. It is inherited by the pure in heart and passed on to advanced initiates who, through aeons of preparation, are ready to present and distribute the Dharma.*

In each new Yuga, or Age, the Dharma appears according to the need of the enmassed souls polarized in certain parts of the earth. The Dharma is selective, falling first into the group-vehicle soul and mind. It is the work of the Dharma to prepare the adept, the initiate and the masses to receive the *Greater* Dharma.

The higher gurus or initiates in certain parts of the East are apprehensive that the Dharma in their part of the world is to be cut off, sensing that those who have carried the Dharma in the East will no longer be politically free to transmit its sacred message. In the Western world, there is now an accumulative conjoining between souls within the inner planes of illumination. Such souls are being prepared as vessels or vehicles for the sacred or seed heritage of a New-Era Dharma light never experienced before in earth.

*Dharma is a Sanskrit word meaning Truth, Law, Virtue, Religion.

In the present era—spoken of as the ending of the Kali-Yuga time or Dark Age of Materiality—the Dharma of the past, suffocated by persecution and assassination, is knocking at the soul-gate of advanced souls who have chosen to be born in the Western hemisphere. How this Dharma is to be used depends upon the ethical astuteness within the minds of those who have established the Jesus Ethic, and whose discriminative faculties will be utilized, enabling them to disburse the sacred organism of the eternal teachings as previously given in the Master-Guru lines.

Approximately two thousand years ago, during His Crucifixion, Jesus shifted the currents of the mental body of the earth, that men might receive the direct Dharma from the Christ.

> *Stewardship in learning of spiritual things is called* Dharma. *To be a guardian of the Dharma —that is, the Good Law or the Truth through Christ—is to truly relate ourselves directly to God.*

To earn the right to receive the Dharma, one must have a sacred attitude toward all pure Teachers and all true Gurus of the Clear Light. This sincere desire will draw one to the source of Dharma-instruction.

Initiation into the Dharma or Dharmic-Light is a holy process supervised by Teachers and Masters, seen and unseen, whose composite soul-power—as a vehicle of instruction—sustains the novice or apprentice within the Dharmic Light.

The West is the crucible for the New-Era Dharma. Egos of the West seeking the spiritual life are probationers and apprentices, that they may take up the yoke of the Dharma and walk in the Light.

The great Masters of the White Line of Gurus in the inner planes, having established the Dharma of the East, are now being drawn toward the soul-light of those in the West who seek to learn, to know and to act.

The Rishi-Wisdom and Ethic line is merging with the Christ Light and the Jesus Ethic, which until now has been little understood in the West and the East. From this will come the New-Era Dharma, the supporting instruction for the forefront souls who must bear the yoke of the Dharma. Their duty is to pass on the Dharma to those who are to inherit the new and vital vision of the golden future when science—having entered its reign of physical ripeness—will be married to Spiritual Reality.

The students of the West, when synchronizing the Dharma of the East with the Christian Ethic as stated by Jesus in the Sermon on the Mount, find their clues to harmonizing the Jesus Ethic with the Eastern illuminative procedures for the spiritual life.

Now coming to birth through the Cosmos Disciples in the West is the Coinciding Dharma. This is the Dharma of the Christ taking one over the duality seas of the microcosmic into the macrocosmic. The Esse under Jesus leads the disciple into dimensions beyond sense into the Divine.

All ethical Teachers teaching from the direct formulas of the Spiritual Worlds ask nothing for themselves, but ask everything for what they carry as the Dharma, as the Light. If one would follow a Teacher with the Dharma, he must be prepared to open with heart expectations what the Teacher brings, offers and provides.

There are different aspects of the Dharma: the Dharma one receives through the Scriptures; the Dharma one receives through meditation; the Dharma one receives directly from his own Teacher or Master; and the Dharma one retains through the direct experience of absorption within the Samadhi state of meditation.

Niscience* has been entrusted with the Western Gate of the Dharma. In Niscience we have the *Versatile Dharma*.

*Niscience is a system of instruction, worship and research founded by the author.

The Dharma flow in the Western consciousness, supported by the Christ, cools the heated fervors in the psychic nature, producing a dimensional rationale suitable to the converging processes between science and religion. Surfacing Dharma now saturated with the Christ will produce a spirituality never seen before in this earth. Upon the West now rests the yoke of this Dharma.

The Masters have scanned the horizons of the soul, and into this world there will come the two or three who will drive the chariots of the Dharma of instruction. Niscience is the opening of the door to these Archetypal-sent Teachers. Niscience is speaking now to the world.

There is the advanced householder who is preparing to become the initiate, and finally, the avatar. He must live in the three-fold action of Initiations of the Dharma. In this, he balances the outer-world demands with the spiritualization of his intellect, and also he adheres to the Family-Atom Altar, to his research of Principle and Ethic, and to his knowing awareness in various stages of Illumination during and after Initiation. He is the threshold-teacher ready to take on the apprenticeship of direct Archetypal training in the Archetypal Dharma. He works in the world as a chela self-acknowledged, as a responsible initiate in placement, whether work, home, family or discipleship. He becomes co-atom to his Teacher and, finally, to Jesus. Wherever he is, he represents the Jesus Ethic and lives it.

> *Omnipresence is ever at hand through the God-sent. By this is the Dharma proved and the Presence of God made plain.*

THE DHARMA IS AT HAND

> *He who lives and practices Truth with love is the life in the Dharma.*

The Dharma until now has failed to unite with the Cosmos. The Dharma for now must rely upon the Cosmos. Those who represent the Dharma must incorporate themselves within the Universal or the Spirit of God which is the Atma holding together everything.

The advanced soul of the West is presently being initiated through three compulsions: (1) The incoming flow of the Dharma from the East into the West; (2) the flow and interflow between realized souls both in this world and the world called Spiritual. These coinciding situations will produce a new era for the awakened soul which, in turn, will keep the pivot or equilibrium point of the equilateral functioning in the human condition. (3) The third aspect of the Dharma is the Archetypal Dharma.

The health of the soul to be wholesome must be functional. When men as souls prepare to become soul-realized, the Dharma comes. The vehicles relaying or distributing the Dharma are rarely recognized in the tumults preceding the downpouring of the Dharma.

Dharma, as applied instruction, rectifies the life of the one who is approaching the climb toward Realization. Dharma, falling from its homeplace in the heavenly reaches, always follows a period of mind darkness. Dharma proceeds out of the provisionary laws supporting the time and the need.

The hour of the Dharma is at hand. Mind chaos is to be vitalized into a total volatized absorption by the mass of men as well as by the individual.

The dark, brackened stagnation of the tamas-energies and the rajasic irritability of the psychic-mental are to be calmed by the sattvic state of soul-infiltration under the Archetypal Light of the Christ.

When will this come? Is it possible in this world for all mankind to be *All-Soul Realized* at one time? The answer is: The All-Soul Realized state is a process and a goal for all men in their inner nature. The *All-Soul Realized* state of total mankind is a timing timed to milleniums ahead.

All men individually are being pressured by the rajasic to the point of desperation. Each soul in the world in his own unique degree of karma is moving toward a calming time of tensions. He seeks to be relaxed to a respite gained by the holy blessing of the *Dharmic expediencies* falling upon men. Men having suffered the fires of hell are to feel the cooling of the snowflakes falling upon their pain.

The Dharma comes first in the smaller inclinations suitable to the receptivity and the call.

The flow and the interflow of the Archetypal Dharma presently is shepherding the Scriptural strengths in all true and proven Spiritual Paths. The Archetypal Dharma is to produce in this closing Kali-Yuga time of darkness a *unitiplicity* of cultures based upon secure spirituality residing in the religiosity inclination of soul-nurtured men.

What men call today *Evangelistic* and *Ecclesiastic* are to be free from creed-restraints. Experimentation and limitation are to be unified and solidified for those on the Religious Path. On the chess board of the soul-game, *all go forward* who seek God. The Dharma is at hand!

Hold fast to the Dharma as your lamp.
Blessed is the teaching of the Dharma.
—*Buddha*

ARCHETYPAL KINGDOM

I knew a man in Christ above fourteen years ago . . . such an one caught up to the third heaven.

—2 Corinthians 12:2

THIRD HEAVEN

Blessed are the poor in spirit: for theirs is the kingdom of heaven.

—St. Matthew 5:3

SECOND HEAVEN

Repent ye, for the kingdom of heaven is at hand.

—St. Matthew 3:2

FIRST HEAVEN

But seek ye first the kingdom of God, and his righteousness; and all these things shall be added unto you.

—St. Matthew 6:33

2

ATOM EVOLVEMENT

Spiritual power is a process of the mind's action in Light out of atom evolvement.

Spiritual Atoms and Meditation

The atom is the key to Eternal-Spirit knowing.

The atom referred to in this volume is not the atom known to the physicist of our time, but is the regenerator within the Eternal Fiat or Eternal sustaining life in God. These atoms constitute the bodies yet invisible to man's physical perception and are the means of his sustaining life after death and of his repeating his lives from experience to experience in earth.

Through re-embodiment, man experiences *atom evolvement*, as it is within the plan of this earth that he always increase his field of action within these atoms. Hence, the atoms are the means of man's bringing more Light into the world. These atoms, yet unknown to man in his outer consciousness mind, are known in his night's sleep and after death. What the bones are to the physical body, these atoms are to the spiritual body.

When research finally confirms that there are *certain impenetrable atoms* holding together the soul-life substances uniting all forms, this knowledge will bring man closer to the ebb and flow of Universal Intelligence. Men will begin to understand birth and death, and will open themselves to the many Heavenly Voices seeking to speak into the deafened ears of materiality.

The dedicated practices of the higher life enable the disciple to free his *spiritual atoms*, which are the permanent portions of Life, Image and Light. The spiritual atoms, which remain eternal in all worlds, are the means of sustaining the physical experiences on earth and the spiritual experience in the World of God. These atoms are not the atoms of force, nor are they organic. They are atoms of the pure stream of the Eternals and consist wholly of Light and of the Body of God. They make up the fabric of man's interweaving and intermeshing; a network of Life, Image and Light, enabling man to function in the world of form and still have entrance into the World of God.

The atom evolvement of each disciple determines his rate of progression. Meditation makes it possible for the disciple to achieve a higher degree of atom evolvement.

True meditation is the greatest power in the overcoming of human karma. True meditation, however, is impossible without atom alignment. When one is out of alignment with his spiritual atoms, his thoughts and feelings during meditation are colored by personal uncertainties and desires. While meditating, the disciple should refrain from analyzing and weighing his personal actions and thoughts.

The chief barrier to perfect meditation is a desire for instantaneous evolvement. The disciple should relate to the Eternals and work with the eternal patience.

When one has mastered the spiritual art of meditation, the Higher Self dwelling within the *Eternal Sustaining Atom* speaks into the soul and into the mind; and one learns of the many processes and facets of the Eternals. He no longer resists the Great Laws governing pain and joy on earth, and understands them to be a necessity in man's emotional and mental evolvement.

When the spiritual atoms are free, the Higher Self is free—and one becomes an heir to the promises of the Beatitudes or Sermon on the Mount as given by Jesus of Nazareth.

Niscience, meaning *knowing beyond knowledge*, is a cognition higher than the instinctual knowledge used by the majority of men in the world. Niscience comes only to him who is ripe or ready. Niscience enables the disciple to relate himself to the World of God and to the world of man.

When the disciple has reached the pure state of Higher Self-Genesis—a state of individualization made possible by many lives of spirituality and dedication—he becomes Niscient or Knowing through atom evolvement; and he uses the higher atoms of his mental body.

Jesus performed His miracles with spiritual-atom power. He used His atoms to rearrange form structure, to extract energy. Where there seemed to be vacuity, He drew forth substances. He exploded His physical body while in the tomb, and thus was translated into His higher etheric or spiritual body. He retained the power to use His physical atoms after His death; thus, He was able to reconstruct His body, that it might be felt by Thomas as being actual flesh.

This is the answer to the mystery of all miracles: one who performs supernatural acts or feats has command over the gravity-atom forces of the earth. This power is used to perform miracles by those who leave the earth. It is a power utilized by the Saints, sacred personages, the Most High Saints, Mary, and all of the Beings who have benignly visited the physical world. Supernatural power is possible for these Beings because they have incorporated into their knowledge the command over the atoms of the physical world.

Whenever a miracle is activated, spiritual-atom knowledge has been manifested. The power over gravity enables one with advanced soul-power to render a service beyond telepathic induction. Manifestation power is at the root of all miraculous intercessions, visitations and healings.

All disciples have lived in many eras and times so that their spiritual atoms might become antennas of sensitivity. Through many lives of countless sacrifices, expressions and evolvements—and always through the love of labor and light—the disciple produces the combining-atoms which unite him with his fellow disciples and the Presences of Heaven.

The first and original pattern for man's evolvement in earth can be found in the atom. If one has the knowledge of the atom, he has the knowledge of the Atma.

The Eternal Atoms are called the Atma Atoms.

The molecule works with the Archetypal Plan and Blueprint to produce the prototypal form and format of life. The total Maya is a dense molecular system, which men call *matter*.

Each phase of the atom correlates to one of the bodies of man. There is first the atom which is the physical side of the atom. Second, is the electron, which relates to the electrical and etheric processes of the etheric planes, the etheric nature, the etheric body. Third, there is the proton, which relates to the astral-emotional and the mental bodies. The neutron relates to the transcendent and spirit. And the molecule action relates to the directing influence of God according to His Archetypal Plan.

The neutrino holds the key to the mystery of the atom.

All is Light. The "I" is a precious particle of Light through which God creates. The "I" is always in a state of working to extract Light from matter. When the "I" is unaware of the Spiritual Self, it works to extract Light from force. Atom evolvement determines the degrees of the "I's" experience.

Each atom in the bodies of man, when quickened in the Light, is as a starry portion of Greater Cosmos. Each atom is a system of Light unto itself. When the atoms within the bodies of man are coordinated to one another, an alignment is made to the Light of the Christ; the body of man becomes a body of Light.

> *The light of the body is the eye: if therefore thine eye be single, thy whole body shall be full of light.*
>
> —*St. Matthew 6:22*

The Atoms and the Bodies

> *I will praise thee; for I am fearfully and wonderfully made: marvelous are thy works; and that my soul knoweth right well.*
>
> —*Psalms 139:14*

Man's spiritual evolvement is yet in the embryonic stage. This earth is the womb for his coming forth. Jesus of Nazareth is the ideal Prototype or Blueprint of the perfect man. Previous to entering this eternity, Jesus had command of the twelve atoms in each of His four bodies.* Thus, He had the power of manifestation and de-manifestation. Through this power He was able to change one substance into another (water into wine), to multiply substance (fishes and loaves), to heal the sick, to change one energy into another, to disappear and appear, to command the elements, to walk

*The four bodies are the physical body, the emotional body, the mental body and the etheric body.

on water, to raise the dead, and to teach of the everlasting Kingdom of God. Being a firsthand witness of the Kingdom of God, He could instruct men as to Heaven's reality.

The higher etheric body or everlasting body of man consists of twelve eternal atoms, which remain unchangeable—retaining their original eternal relation to one another. The average man must work for aeons to perfect the atoms in his physical body, emotional body and mental body.

In the present age, man's physical body contains nine orifice atoms, correlating to the nine orifices or openings of the physical body.

The emotional-body atoms of man are now the determining factor as to his evolvement. Some in earth have evolved only the five lesser or sentient atoms of the emotional body; others have evolved seven, and, in some rare instances, ten. In the emotional body may always be found one atom which is *individualistic*. During certain intervals, this Individualistic Atom becomes an agitator to the atoms within the physical, emotional and mental bodies. Through this agitation, man is initiated into a greater degree of consciousness.

If a person has but five sentient atoms activated in his emotional body, the agitation caused by the Individualistic Atom will result in some form of violence to, or suffering in, the physical body; some form of emotional tension in the emotional body; and some form of disturbance in the thoughts. Such pain and suffering refines the senses and makes them better instruments for the soul. If a person has activated seven emotional-body atoms, the Individualistic Atom works upon the physical, emotional and mental bodies to make each body a better vehicle or instrument so that one may work more directly with his soul. If a person has activated ten emotional-body atoms, the Individualistic Atom produces a heavenly recognition of him as a personage or a holy presence in Heaven and on earth. The Individualistic Atom in the emotional body of the holy person enables him to return to the earth if he so desires, or if there is need for his presence on earth.

Until man has activated twelve atoms in each body, the Individualistic Atom will continue to agitate, stimulate and stir the other atoms in the various bodies. The purpose of the Individualistic Atom is to make man aware of his eternal identity and of his soul and Heaven. As long as the Individualistic Atom is active, there is heavenly hope that man will seek a way out of materialistic and sensual expression.

Each person, upon entry into this eternity system, quickened three mental atoms in his mental body. Regardless of the degree of his evolvement, each person of this earth has three mental-body atoms: (1) the Supreme-Will Atom; (2) the Imaging Atom; and (3) the Eternal-Light Atom.

The *Supreme-Will Atom* of the mental body relates man to the Will of God. The *Imaging Atom* of the mental body relates him to the Father. The *Eternal-Light Atom* of the mental body relates him to the Christ and to the Greater Archetypes under command of the Christ.

Through many initiations and spiritual advancements, some persons will gain mastery of the three mental-body atoms, and thus will free the nine dormant atoms of the mental body. Others of the earth, being in love with the physical life and their physical selves, will be unable to coordinate and master the three mental-body atoms; when this earth eternity system has concluded, such persons will be compelled to undergo initiations in other eternity systems until they have total command of the three mental-body atoms.

Until man has total command of the Supreme-Will Atom of the mental body, and is at one with the Will of God; until he has mastered the power of imaging selflessly; until he has the power of revelation through the Eternal-Light Atom, relating him to the Christ—his thoughts will be concerned more with his self-interest, his emotions will be used for self-preservation, and his creative works will be concerned solely with self-aggrandizement in the world. When the Supreme-Will Atom, the Imaging Atom and the Eternal-Light Atom are in perfect alignment, man will no longer seek to preserve himself; he will work directly with the Will of God, the Imaging Power of the Father and the Archetypal Light of the Christ.

One can best determine his own degree of mental evolvement by observing the predominant theme of his thinking. If his thoughts are saturated with self-will and the desire to gain for himself, he is using the lower aspect of the Supreme-Will Atom of the mental body; but if his thoughts are filled with the will to do the good, the pure and the real, he is preparing to come under the guidance of the Will of God.

If one is using the slower rhythms of the Imaging Atom, he will think traditional thoughts; he will be satisfied to conform to the thoughts of the masses. However, if he is highly evolved as a vessel for humanity, he will become at one with the Father, and the pure stream of imaging will enter into his thoughts.

If one desires truth and has the courage to envision the new spiritual transitions, he is influenced by his Eternal-Light Atom correlating to the Christ. If one has an inquiring mind and lacks reverence for the minds of others, he will appoint himself as a messiah over others. But if he has gained the spiritual insight throughout the ages, he will pursue Light as a spiritual neophyte or acolyte; he will aspire to be illumined in Christ. The Eternal-Light Atom will unite him with the Greater Archetypes and he will come under the tutelage of the Christ Spirit.

When a holy person thinks, the Supreme-Will Atom, the

Imaging Atom and the Eternal-Light Atom are coordinated —and his thoughts are creative, fulfilling the perfect equation in thought.

The atoms within the higher etheric body enable the atoms in the physical body, the emotional body and the mental body to penetrate, to fortify and to influence one another.

In each body there are progressive atoms, or atoms which have been quickened throughout the ages into progressive states. These atoms work more freely with the higher etheric body. Until the time of Jesus, the progressive atoms were used for survival. The Resurrection of Jesus began a quickening of a higher degree of atom progression in the humanities. After the coming of Jesus, those who truly followed His precepts experienced a quickening within the *Indestructible Atom* in their foreheads. Such enlightened ones began to stir other atoms within their bodies and to learn more about their spiritual natures and the Spiritual Worlds.

Many in the world are yet content to express sensuality through the five sentient atoms in the lesser emotional body. When one has reached spiritual maturity, his sentient atoms no longer command him; the sentient atoms are stilled within the lesser emotional body—and his soul faculties are freed. He produces a quickened intuition, rather than instinctual sensing. His appetites become more selective, his charity more expansive. His eye beholds the souls of men; he hears the greater harmonies of the Universe. He loves all things and all creatures of the earth, and he devotes himself to God with his heart, mind and soul.

Prayer makes pure the oil in the lamps of the senses, and the soul-faculties are free to give forth a pure light. Through meditation, the senses become the servants for the soul, fulfilling Jesus' parable of "ten virgins" (St. Matthew 25:1-3). The five sentient atoms in the lesser emotional body may be likened to the five foolish virgins; the soul-faculties working through the five higher emotional-body atoms may be compared to the five wise virgins.

The spiritual hope for man is that the eternal work of God is endless within him. One by one, all atoms in their progressive, spiraling state will be set into alignment with the atoms of the higher etheric body, and therefore into the perfect light.

The planets, the Earth and the Sun of this eternity system form a mathematical polarity through which the Spirit of God works. The Sun, the Earth and the accompanying planets are encased in a mighty spiritual atom, called the *World-Soul Atom* or *Diamond of the Ages.*

Each thing in the earth—from the plant to man—has an *Eternal Sustaining Atom.* Each Eternal Sustaining Atom has a degree of light correlating to the Will, Life, Light and Love of God. The degree of light within an Eternal Sustaining Atom determines the form it will manifest and express, and also determines what it will contribute to the universal plan.

In the beginning of this eternity the Elohim-Hierarchy dwelling within adjacent constellations, and our Father of this eternity, sounded Life-Tones into the World-Soul Atom. Their combined Life-Tones played upon the innumerable Eternal Sustaining Atoms dwelling in the cosmic mist of the yet unformed earth. As the earth became a coagulated body or mass, the four latent bodies within the Eternal Sustaining Atom of man began to shape in vapor-like forms.

In man's Eternal Sustaining Atom there is a soul-pulsation. The soul's pulsation, working with the Life-Tones, quickened the sentient atoms within the emotional body. Next, nine physical-body atoms were quickened; and following this, three mental-body atoms were quickened.

The Life-Tones of Hierarchy and the Father will continue to play upon the bodies of men until men have reached a state of perfection in this earth. As men evolve, the latent atoms in their various bodies will come forth.

With the coming of Jesus, the Christ Light-Tones began for this eternity. When the Christ Light-Tones penetrated the earth system, there began a coalescing between the Life-Tones of the Father and Hierarchy and the Light-Tones of the Christ. When the Life-Tones and the Light-Tones converged, man began his long upward climb toward a perfected mentality. As the Light-Tones of the Christ continue to move upon men, men will become more and more noble in their thoughts, and their minds will become vortices of creation.

In *Cosmos-Genesis*, man will have command of his emotional-body atoms and perfect control of his mental atoms.

THE ETERNAL SUSTAINING ATOM AND THE HIGHER SELF

> *The Central Atom in the Spiritual Worlds is the Eternal Sustaining Atom. Each man has an Eternal Sustaining Atom. The Higher or Eternal Self resides within the Eternal Sustaining Atom.*
>
> *When man entered this earth or eternity, he was etherically encased within the Eternal Sustaining Atom. This Atom has the appearance of a great diamond.*

The Eternal Sustaining Atom sustains life eternal from one eternity to another. The Eternal Sustaining Atom is the means through which one moves from one eternity to another. When one prepares to enter into a new eternity or the beginnings of a new eternity, his soul and the atoms for his future bodies are encased etherically within the Eternal Sustaining Atom. Even as a person is unaware of his encasement within the womb of his mother, so was he unaware of his Eternal-Sustaining-Atom encasement when he entered this eternity. The Eternal Sustaining Atom works with God's Creative Fiat and adapts itself to the eternity in which it finds itself.

The Sun, the Earth and the accompanying planets are part of a network of *eternal vertebrae* consisting of eleven other eternity systems. When these eleven sister and brother eternity systems reach a certain balance in their equation, they merge their Tones with the Alpha-Tones as sent forth from God. These Tones become mighty transmitting streams of energy and light through which the various Eternal Sustaining Atoms containing the forms of life-to-be are transported into the void of this eternity system.

In the first Edenic or etheric intervals of the Earth's creation, the Father of this eternity—working with the Elohim-Hierarchy, the Archangels, Angels and other Presences—imaged all things as they were to be formed in this eternity. The Father, using the Life-Fiat, animated the original Blueprints or Archetypes of all living things to come.

The Eternal Sustaining Atom is the supreme atom of the eternal atoms of man. When men were first immersed and encased in this eternity system, their eternal or spiritual atoms—containing aeonic intelligence—began their responding immediately to the cosmic pre-natal helps given by Hierarchy, our Father, the Christ and the Archangels.

The word *soul* comes from sol or sun. In the first Great Intervals of this earth's creation—as man moved through the Moving Deep, and while the Sun was yet in a nebular state—the Eternal Sustaining Atom, the Higher Self and the soul received their quickening for the Earth's work.

The Archangels make the bridge of light into the Universe.

The Archangels make it possible for souls to flow into other eternity systems. They provide an energy passage-way between eternity systems. The Archangels are path-builders, using great rays of light as suction, that souls may enter into new eternity systems awaiting their coming.

The Eternal Sustaining Atom within each soul is obedient to this thrust between eternities. In each Eternal Sustaining Atom is an Archetype and Prototype of what man is to be in the eternity system he is to inhabit.

As semen in the male of the Earth and ovum of the woman produce an embryo, so the Eternal Sustaining Atom contains the potential likeness or image of what it is to be in an eternity system.

All soul-journeys into new Earth-systems begin when the eternity system they dwell in is in a dying state. When the timing is equipoled for a soul-journey to another eternity system, that which is to be the life-form expression in a new eternity system sets out encased in an Eternal Sustaining Atom. The Eternal Sustaining Atom unfolds simultaneously with the progressive gravity life of the new eternity system.

Man never disconnects himself from his Eternal Sustaining Atom which overdirects what sort of form he expressed from life to life.

The more man is developed as an ego in Earth, and believes the ego to be everything, the more he obscures by a defaulting energy his knowledge and experience with his Eternal Sustaining Atom, where dwells his Real Self. Thus, ego-bondage seals away man's knowledge of God as Eternal Spirit in him. On being initiated into one's true nature, Eternal Spirit—remaining always supreme yet obscured—speaks to him and tells him of his reality.

All that the soul has experienced in former eternities, all that has been earned, is encapsuled and retained in the Eternal Sustaining Atom.

Light is the very essence of man's eternal being. Each soul on earth attained certain degrees of light while living in other eternity systems before this earth or eternity system was ever created. The degrees of light one attained in former eternities are sustained eternally in the Eternal Sustaining Atom; they are never lost. The eternal degrees of light work in unison with the pulsation of the soul.

When the soul enters a new eternity system, the Cosmos Atom of the new eternity quickens the soul's pulsation within the Eternal Sustaining Atom. Within each Eternal Sustaining Atom there is also a Cosmos Pulsation enabling man to work with the Greater Alternates of Cosmos and to respond to the Eternal Laws. The creative side of the *Cosmos Alternates* is used by Hierarchy, the Archangels, the Christ and the Father to shape and form men.

The Eternal Sustaining Atom enables the soul to work with two polarities. This causes a duality action of positive and negative in the emotions, thoughts, conscience and actions of man. This duality action also enables one to take either a masculine or feminine form, and provides the way through which one may work in the physical gravity world and still remain communicable to the Heaven Worlds. The Eternal Sustaining Atom will make it possible eventually for man to give birth to a third or triplicity polarity action; this will produce an androgynous will and mentality.

In this eternity, the Eternal Sustaining Atom is at home within the Archetypes of the Kingdom of God. When one uses the multiplicity power of his spiritual atoms, the Eternal Sustaining Atom and its eternal action will be disclosed to him. And there will no longer be any doubt as to eternal life and eternal being.

The Higher Self, dwelling within the Eternal Sustaining Atom, is the product of other eternities. Through the aeons of this eternity, the Higher Self, working with the Eternal Sustaining Atom, projected one body after another. The Higher Etheric Body was the first body to be projected in this eternity. The Higher Etheric Body, the Higher Self and the Eternal Sustaining Atom, working with the soul, are in command of all of man's various degrees of evolvement and development.

All good works of former eternities adhere to and dwell around the Eternal Sustaining Atom. This Eternal Medallion of Grace is called *the Diamond Medallion* because of its scintillating brilliance.

The Higher Self is located directly over the head. Through the process of many lives or re-embodiment, man shall bring forth the greater atoms in all of his bodies so that he may eventually be at one with his Higher Self. And he shall become both the Alpha and Omega, the first and the last.

The Higher Self or *Diamond of the Self* sounding the Music of God is seated in the Eternal Sustaining Atom. This Eternal Atom assures man of his soul life and of his physical life. The soul-tone sounds from the record of grace or karma determine whether one's life will be harmonious or inharmonious. One cannot escape from the tone or sound creating his destiny. It is uniquely his own, moving him forward through the infinite and varied processes of creation.

The basic container of the soul-tone is the Eternal Sustaining Atom, centered approximately three feet over the head. The tone within the Eternal Sustaining Atom is sounded by Eternal Spirit into the soul's medallion, which is a superconscious matrix of pulsating light and tone. This soul matrix or medallion interpenetrates and covers the brain, the forehead, the ears, the nostrils, and radiates outward into the aura of each person, forming a nimbus field of supercharged light.

When in alignment with one's own soul-tone, one may at first receive many variations of his soul-tone. From the basic tone or master tone within the Eternal Sustaining

Atom, one gains the power of universal consciousness and is then free from the darkened locked-in gravity tones of Maya or limited consciousness.

The sounding of the *Om* before or during meditation energizes the soul's medallion, the soul-tone, and the Eternal-Sustaining-Atom tone.

The *Galaxy Consciousness* begins at the point of the Higher Self and goes to the point between the eyebrows. Between the Higher Self and the eyebrow center are millions of atoms in direct communication with the Galaxy Fatherhood of our eternity system and of our eternal origins. When one is in communion with the Galaxy Atoms pouring down from the Higher Self, he is at one with the Total Universe.

The uncreated is always trying to become the created. When one reaches the Galaxy Atoms, he is in the Transcendent. The Eternal Sustaining Atom is like a great generating system when the holy interconnection is made between man and God.

The Universal religion for the coming mankind will be supported by scientific confirmation that man is more than ego. All advanced seers and initiates within the Avatar light will work in the New Era through the use of an *Omniscient Cell* in the brain. The opening of the Twelfth Chakra or the Eternal Sustaining Atom of man will provide the brain with an Omniscient Cell which will give the revelatory powers of Cosmos Prophecy. Through the use of the Omniscient Cell, the spiritual seer will work as a cosmos prophet with Christly revelatory powers.

> *The King of Kings came to this world to give to men a crownship or a royalty. Each one of the earth is a potential son of God, a soul existing in an earth awaiting the true birth and fulfillment.*

THE SACRED ATOM OF THE HEART

There must be atom evolvement within the Heart Chakra before the conscience can function. The conscience cannot and does not function until the Sacred Heart-Atom is opened.

As the stars are the atoms in the Body of God, each Eternal Sustaining Atom is a cell in the Body of God. The Higher Self, dwelling within the Eternal Sustaining Atom of man, projects from itself the many and varied prototypal expressions.

The Eternal Sustaining Atom, in the forming of the various prototypal bodies, is assisted by the Indestructible Atom between the brows and the Sacred Atom within the heart. These three atoms work in a triad action so that man will remain an eternal being while experiencing the prototypal works in the earth.

The Sacred Atom of the heart is deathless. This Atom holds a microscopic etheric form which records all physical actions and desires. When death comes, the record of the physical life is retained within the Sacred Atom of the heart. In each life, the same Sacred Atom returns to rest in the physical heart.

There are 24 corridors or chambers within the Sacred Atom of the heart. These correlate to the 24 Elders spoken of in the Book of Revelation 4:4; they also correlate to the 24 Divine chromosomes affecting man's Divine ancestry, rather than his progenitorial line or ancestry.

All forms of healing, physical or spiritual, are mediative. However, no healing is possible when the Sacred Atom of the heart and the Pulsation of the Soul are separated by lack of faith. The Sacred Atom of the heart contains the restoring-life sealed into all men by the Father since the beginning of this eternity. When one offends the Sacred Atom of the heart by offensive and self-willed actions; when one abuses love, or follows shallow intellectual pursuits—the Sacred Atom of the heart, and the restoring-life dwelling therein, fail to give forth the renewing and regenerating vitality. Illness, sickness and discomfort are the result.

When one has faith "as a grain of mustard seed," he will receive a healing grace from the higher powers of Mediation. Such faith enables one to receive a permanent healing. When one has faith, he enters into the greatest mediative arteries of God. Faith enables the Heavenly Helps of Mediation to manifest their perfect works, and thus unite the sufferer with the restoring life sealed into the Sacred Atom of the heart by "our Father which art in Heaven."

> *The working of the mind bringeth to men the coming poise within the next 1,000 years. The bodies of men shall bring anew the heart muscle, on which the life-light within the Sacred Atom shall stand poised and burn within the Master's Light, the Light of Christ, the Hierarch's Fire onto God.*

The spiritual fire within the spine seeks to unite with the heart, the throat, the crown of the head. If one has loved sensually, the spinal fire in the heart is congested, painful. When one loves with purity, the spiritual fire has free ingress to the heart. The Sacred Atom becomes unobstructed; the heart becomes a lighted chalice for Divine love.

The power of luminosity or of disappearance is gained through higher degrees of initiation. There are four atoms around the heart called *the Luminosity Atoms*. When these are activated, one may appear or disappear at will. Few in the world have attained this power. The Lord Jesus and the men of the Elect used this power to make invisible their physical bodies—to disappear and to reappear. Elijah used this power to rise into the Heavens. Enoch translated his body with this power. Jesus self-cremated His physical body through the activation of the Luminosity Atoms around the heart, and was thus resurrected.

The physical man hath a body with a heart.
The spiritual man hath a heart with a body.

THE INDESTRUCTIBLE ATOM

Until the coming of Jesus, the master bliss-center in the middle of the forehead was unavailable to Kundalini flow. Since the fulfillment of the work of Jesus, this master bliss-station is open. It is called the Indestructible Atom.

The Christ Spirit through Jesus is quickening the Indestructible Atom in the center of the forehead of those who would serve God.

The Indestructible Atom, a tributary for the Christ Mind, sets the tone for what is received by the highly-evolved disciple.

The soul is the mediator between the Eternal Sustaining Atom and the Indestructible Atom, where dwells the ego or individuality of man. When the Sacred Atom in the heart, the Indestructible Atom between the brows and the Eternal Sustaining Atom are at one, man will attain the power of manifestation and de-manifestation, even as did the Lord Jesus. Jesus of Nazareth had the power of manifestation and de-manifestation; He was able to change water to wine, heal the sick, and raise the dead because the Sacred Atom of His heart, the Indestructible Atom between His brows and His Eternal Sustaining Atom worked as one.

The Christ-Spirit quickened the Indestructible Atom of

man. Due to the quickening and the emergence of the Indestructible Atom, the mentality of man has undergone a tremendous transition within the last 2,000 years.

The Indestructible Atom enables man to receive the Power of the Holy Ghost. The Power of the Holy Ghost works through Tone, Light, sound and thought. Before the coming of Jesus, the Holy Ghost was experienced through the World-Soul Atom and the Race-Lord Jehovah. Since the coming of Jesus, the various degrees of the Holy Ghost are experienced through the quickening of the Indestructible Atom.

The Indestructible Atom will eventually enable man to image with manifestation powers. The lower aspect of the Indestructible Atom works with a de-manifesting action. The highest aspect of the Indestructible Atom works with a manifesting action.

> *He who blendeth the Indestructible Atom within his forehead with the Medallion of his Soul dissolveth his past, bringeth forth the love, and standeth before the Door with the key in his hand.*

There are seven actions of the Indestructible Atom. The first action of the Indestructible Atom is de-manifestation. The instant a person turns toward the spiritual life and begins the practice of dedicated meditation, the Indestructible Atom is quickened and stirred. Being a novice in meditation, he has yet to enter into the portals of the Eternal Sustaining Atom, where dwells the Higher or Sacred Self. Therefore, he will undergo a necessary de-manifesting and dissolution process, which consists of an extreme cleansing of the negative personality traits. In this, old forms are dissolved; habits are changed; thoughts are clarified and organized; and concepts are corrected.

The second action of the Indestructible Atom is to dissolve the *prototypal shells* carried over from former lives, thereby enabling one to receive the lessons and knowledge gained from the Dweller Initiations. Thus, he undergoes a dying to the personalities of former lives so that he may be born to a luminous personality and individuality within this life.

The third action of the Indestructible Atom is to aid the disciple to correlate with the Angels of his bodies so that he may rise above the Astral World and begin his Night-Serving. When one masters the Astral World, he has mastered sound; and therefore he opens the inner ear.

The fourth action of the Indestructible Atom is to enable the disciple to translate his Soul-Grace into outer works in the physical world. (The Indestructible Atom is not one of the three Mental Triad Atoms.)

The fifth action of the Indestructible Atom is to familiarize the disciple with the Hum of the Archetypal Thread and its work with the spinal cord, the Sacred Atom of the Heart, the Logos Atom of the throat, the Pulsation of the Soul, and the Eternal Sustaining Atom.

The sixth action of the Indestructible Atom relates to initiation within the Higher Worlds, inclusive of the sacred regions of the Astral World, the Spheres of Light and the Realm of Light. When the fire within the spinal cord, the magnetism within the silver cord, the tone within the Indestructible Atom and the Hum within the Archetypal Thread intermingle and blend, one may sustain his action in the Spheres of Light and the Realm of Light, and thus receive instruction within the Higher Worlds.

The seventh action of the Indestructible Atom relates the disciple and initiate to the Tone residing within the Greater Archetypes of the Archetypal Worlds, thereby enabling him to receive the Revelation Power of the Holy Ghost. When one has the power to enter into the Great Silence or the World of Tone where reside the Greater Archetypes, he receives the Power of the Holy Ghost; and the Laws of Creation and the Laws pertaining to the Will of God are revealed to him. He becomes a Law-giver and a Law-revealer.

> *The jewel in the Forehead, the Jewel in the Lotus, the shining Light in Christ, the true Light in God—all dwell in the prophet. He who carrieth the Light feareth not in his speaking, and giveth of himself that the Light might come.*

The Jewel in the Forehead is the Indestructible Atom, enabling the prophet to envision or to foresee. The Jewel in the Lotus is the Gleaming Brain or the thirteen radiant portions of the brain surrounding the pineal gland.

Since the coming of the Christ, the Indestructible Atom has placed extreme pressure on both the pineal and pituitary glands. This has created extensions of self-aggression in mankind at large. This also has extended the Genesis stages in evolvement. The present Self-Genesis coming to birth in mankind is creating disunion rather than union.

> *The word* Wisdom *cometh to man through the heart. The word* Light *cometh to man through the Divine Atom in his forehead. The Divine Atom extendeth its light to the gland between the eyes and illumineth the spiritual crown within the head.*

The Divine Atom between the eyes, in the brow, is the Indestructible Atom. Since the coming of Jesus, the Indestructible Atom has been quickened and men are now becoming more individualized, and therefore less primitively bound to ancestral claims.

The Indestructible Atom of the initiate is set to the Central Atom in the Spiritual Worlds and the Central Atom in Earth.

Lord Jesus Christ opened the Sattvic Hierarchy-Nature Gate centered in the Indestructible Atom. Calling on the Name of Jesus opens this Gate and moves one toward the *thirteenth* aspect of his hierarchy nature patterned after Jesus, who represented the *thirteenth* principle among His *twelve*.

Jesus was the Sattvic Presence of God in the midst of the suffering of man. He functioned and operated through the hierarchy nature within the *thirteenth* principle. All hierarchy-image forms created in the likeness of the Father and Hierarchy express themselves through *twelve* Prototypal forms. These *twelve* in the earth in the spirit of goodness seek to become in the likeness of Jesus and thereby express the *thirteenth* principle or their sattvic natures.

The Indestructible Atom works as a Pulsar.

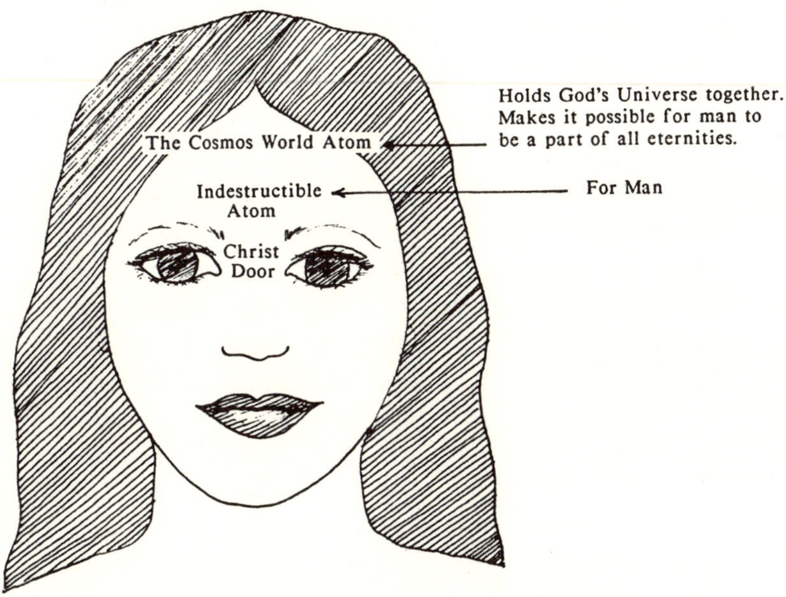

THE ATMA AND THE UNIVERSE

On making total surrender to God and to His Purusha Consciousness which occupies the Indestructible Atom Center, called the *Holy Cave*, one moves beyond imagination and fantasy into direct experience of God-Realization as consciousness.

It has been the custom for ages to begin Kundalini-yoga practices by concentrating upon the base-of-the-spine or Muladhara Chakra. The Western initiate begins at the Anahata Chakra or the Heart Center, that he might go beyond the seven phenomenal or relativity Chakras into the five upward Superconsciousness Chakras. Before the Christ Spirit through Jesus opened the Indestructible Atom in the center of the forehead, or Bliss Whirling Cave of Transcendental Light, initiates in former ages depended upon the seven relativity Chakras; thus, they experienced Samadhi with partial articulateness. In the present time, all initiates —regardless of their religious placement—are working for the high calling of Christ. When this is accomplished, advanced initiates will express and articulate the true Logos or Samadhi speech of Light.

> *The hierarchy station for the use of the hierarchy nature is at the Buddhi Point, or the Indestructible Atom, which is in the middle of the forehead. When one uses the Niscience method of healing, he is using his own hierarchy nature within the Will of God to heal. In this moment of healing, he gives a* Passing *to those in need of healing. The Will of God takes the need into the command of God, that the situation or person one would heal will receive instantaneous correction or healing. In the Niscience techniques of healing, miracles occur, which are in reality the Will of God at work.*

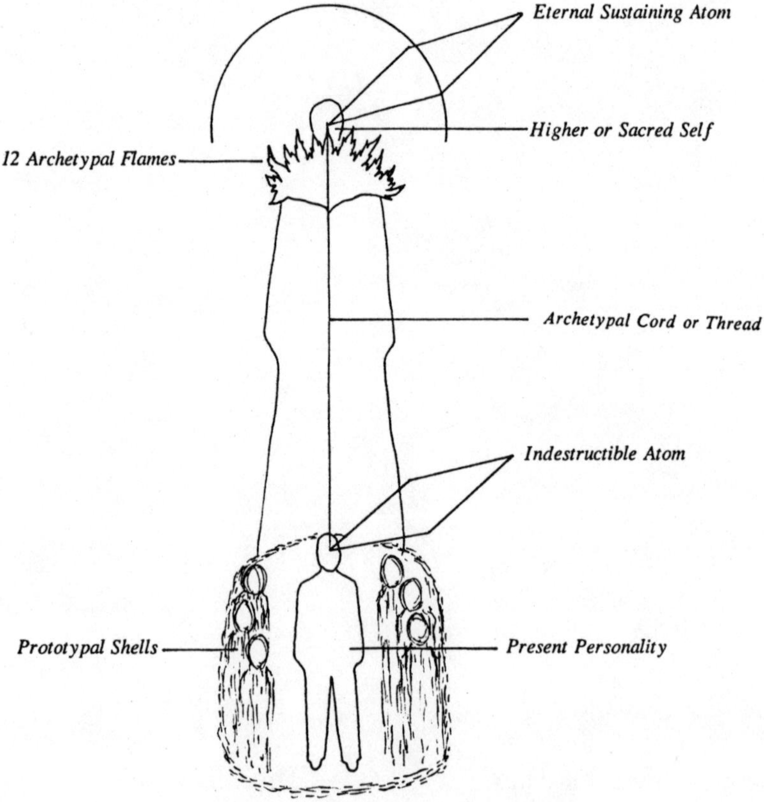

THE ARCHETYPAL CORD

3

THE ARCHETYPAL CORD

Or ever the silver cord be loosed . . . Then shall the dust return to the earth as it was: and the spirit shall return unto God who gave it.

—*Ecclesiastes 12:6,7*

There are three cords or threads of life: the umbilical cord, the silver cord and the Archetypal Cord. The umbilical cord sustains the life of the embryo during the period of gestation and pre-birth. The silver cord or soul's cord enables man to release his everlasting body from his physical body at night during sleep, and it is also the means by which he withdraws from the physical body at death. The Archetypal Cord or Spiritual Cord determines one's response to Spirit and works with the soul to determine the length of life.

In death, the silver cord or soul-cord is severed in three 24-hour intervals. The Archetypal Thread is the Cord between the Higher Self and the soul.

When the Archetypal Cord ceases to sound its Hum, death begins in the physical body and in the lesser etheric body. After death, each person undergoes a period of time in the Archives of Introspection and Retrospection. The records of former lives are read and one enters into rectification as required by the soul after death.

In the highly-evolved person, the silver cord or soul-thread becomes inactive after the records of the soul have been read. The Archetypal Cord takes command; and the one who has died unites himself with the higher purpose of the afterlife. One who is less evolved continues to function through the silver cord and enters into a purgatorial state of rectification.

The silver cord, consisting of a veil-like magnetic and etheric substance, has five antenna points. These points are correlated to the spleen, the liver, the heart, the throat and the crown of the head. The mood for approaching death begins in man by the relaxing of the silver cord. During the first three days after death, the silver cord is loosened and withdrawn from its five antenna points.

The Archetypal Cord or Thread is the connecting link between Spirit and soul. This Cord has a heavenly consistency and vibrancy called a *Hum.* The Hum of the Archetypal Cord sustains the true identity of the spiritual self. When the Hum of the Archetypal Cord decreases, the silver cord loosens its meshlike hold upon the spleen and slackens its attachment to the liver; and the silver cord's tendrils, upholding the flame of life within the heart, are relaxed. With the decrease of the Archetypal Hum, the physical body begins to disintegrate, and the person, regardless of age, enters a dying state.

The soul's record of past lives and the record of deeds in the present life determine the length of life and the timing of death. The Archetypal Cord, working with the record of the soul, withdraws its Hum when the purpose of one's life is completed. Whether one dies as an infant or in full maturity, he has fulfilled God's Equation in some manner. Even though the promise of one's expression may appear to be incomplete, the person has answered and responded to the demands of the soul's record.

To every thing there is a season, and a time to every purpose under the heaven: A time to be born, and a time to die.

—Ecclesiastes 3:1,2

The Archetypal Cord or Thread is a scintillating etheric cable between the Eternal Sustaining Atom, the Higher Self, the twelve Archetypal Flames, the Soul and the Indestructible Atom. The Archetypal Cord continually transmits the heavenly refreshings to the soul and the resuscitating impulses to the physical body.

The umbilical cord, enabling the mother and embryo to be communicable to one another, is one of the projections of the Archetypal Cord. The Archetypal Cord of the person to be born enables the embryo to establish the intricate workings of the spinal cord and nervous system, which in life permit the expression of the personality-will and, eventually, the individuality-will.

The silver cord is another projection of the Archetypal Thread. The silver cord, abiding within the vicinity of the spleen, the liver, the heart, the larynx and the crown of the head, enables a person to interrelate his four bodies and to sustain his travel during dream or sleep. The silver cord works through the spleen to enable man to activate the twelve prototypal reflections of Hierarchy into the billions of blood cells in the physical body. The silver cord works through the liver to enable man to command his muscular system and to stimulate the emotional-will. In a hammock-like fashion, the silver cord supports the miniature living-image within the Sacred Atom of the heart. The next phase of the silver cord's action takes place in the larynx, where the silver cord is the connecting link between the spoken word upon the lips of man and the creative Word of God.

The silver cord and the Archetypal Cord unite within the

Indestructible Atom. The Indestructible Atom, the Archetypal Thread, the silver cord, the Pulsation of the Soul and the Sacred Atom of the heart work conjointly, and determine the end of life or the time of death. When the Indestructible Atom ceases to sound its resonant tone for the present life, the silver cord is "loosed," or that is, released from the physical body and lesser etheric body—and death is inevitable.

With repeated alignment in meditation and dedication, one may become aware of the Hum of the Archetypal Cord. This Hum may be first discerned in the heart during meditation as a feeling of absolute freedom in the sending and spreading of love.

When one speaks a mantram, the Hum of the Archetypal Cord within the larynx permeates the mantramic phrases and becomes the creative word, overcoming and resolving the banked negations gathered within the lesser etheric body of oneself or within the lesser etheric body of the one to whom the healing word is sent. (e.g. Angel to Angel Mantrams.)

When one stills his thoughts in meditation, the Archetypal Cord's Hum overpowers the vibratory hum around the Soul's Medallion—and one then enters into the threshold of the soul's pulsation. From this point, the disciple is penetrable to the Hum within one or more of the Archetypal Flames surrounding the Higher Self. If the twelve Archetypal Flames are penetrated, the disciple is penetrable to the Archetypal Cord's Hum within the Eternal Sustaining Atom. He thus becomes communicable to the Higher Self, and thereby receives instruction from the greater Tone or Hum in the Archetypal Kingdoms. In this manner—and only in this manner—man receives the true Word of God.

The *OM*, which is the sacred sound of the Word, is the

mighty Mantra inviting God-Realization. Due to the submerged spirituality of many souls in the West, there are many Mantra-sounds which Western or Anglo-Saxon logos is yet to experience. It is the work of the initiate to attract to himself his own Naming-of-God Mantra. His Living Teacher assists in this, as it is a gift from the Teacher to his student to give the Holy-Name Sound of God, and thus free the Archetypal Hum in the soul of his disciple.

Soul-movement is sustained, maintained and repaired by the Purusha diamond-like Deva substance. The Purusha-stuff of God is used by the Devas to sustain the cable or Archetypal Cord in the spiraling rise beyond cause and effect, or Maya-repetition.

In the accomplishment of Samadhi through meditation, the Archetypal Cord is drawn down into the Divine Eye within the skull. Through this, one experiences the true and lasting state of Illumination or the *Christ Samadhi*.

The Western disciple and initiate now works toward this ultimate state. The Samadhi meditation of the New Era will no longer be inarticulate or indescribable, but will be accompanied by a descriptive Universal Logos in alignment with the true Voice of Cosmos.

The ultimate in meditation is Christ Samadhi. The Christ Samadhi is different from the Rishi Samadhi. Christ Samadhi is articulation of Superconscious Light and identification of the dimensional planes within the range of Superconscious experience.

There are two stages of Samadhi. Most gurus experience a secondary Samadhi; more advanced gurus reach the deeper Samadhi.

Samadhi is a seeing and observing without the participation of intellect. A disciple of the Christ works to articulate Samadhi and to describe higher planes of consciousness. Only through union with the Christ can this be done.

Many forms of life and many forms of death are recorded in the interior planes of the Archives, but the pure stream in the Archives carrieth the Archetypal Thread onto the Greater Archetypes and God.

THE ARCHETYPAL FLAMES

There are twelve Archetypal Flames surrounding the Eternal Sustaining Atom. The twelve Archetypal Flames make it possible for man to assume from life to life a different identity or prototype. Thus, regardless of what Light-Stream one expresses, he may be an Aries in one life, a Leo in another life, etc. This is God's way of giving man the opportunity to incorporate all of the experiences necessary for his evolvement and perfection. When one finally incorporates the total essences and substances provided by the Archetypal Flames, he becomes a composite identity correlating to the Perfect Identity or the Lord Jesus.

There is a central or thirteenth Archetypal Flame dwelling *within* the Eternal Sustaining Atom. The thirteenth Archetypal Flame assures man of Eternal Life. Until the coming of Jesus, the average person had access to only ten Archetypal Flames; thus, he was not responsive to the Higher Self and was devoid of knowledge of the Higher Worlds. The Passion of Jesus, His Resurrection, and the Christ-Spirit establishing Himself as Sovereign of the earth, activated the Indestructible Atom in the brow of man, thereby giving him the opportunity to avail himself of the thirteen Archetypal Flames.

The Archetypal Cord or Thread is an etheric cable consisting of thirteen delicate, minor threads. Each of these threads has its own particular tone. These tones make up the Hum of the Archetypal Cord. Each thread and its tone correlates to and responds to one of the Archetypal Flames dwelling within the vicinity of the Eternal Sustaining Atom.

When one believes not on God, and is yet overshadowed by the guilts of wrong-doing, the vibratory hum of the soul becomes an impassable barrier; the eleventh, twelfth and thirteenth tones within the Archetypal threads are muted, and the eleventh, twelfth and thirteenth Archetypal Flames are veiled away. Until one activates the tones in the Archetypal Thread that correlate to the eleventh, twelfth and thirteenth Archetypal Flames, he is subjected to repeated lives through Tribal-Genesis and Family-Genesis.

Ten Archetypal Flames correlate to the ten Hierarchs overdwelling the earth. The remaining three Archetypal Flames correlate to the Father, the Christ, to God.

During the time of Jesus on earth, there were some who were ready to respond to the eleventh Archetypal Flame. From these came the early Christians. There were a few, or a small portion, who recognized the Christ in Jesus. Such persons had activated the twelfth Archetypal Flame. Among these were Jesus' Disciples who, upon the death of Jesus and their experience in the Upper Chamber, activated the thirteenth Archetypal Flame, and thus received the Holy Ghost.

When a person seeks to live purely and reverently, the Archetypal threads are quickened, and the Archetypal Flames around the Eternal Sustaining Atom become as radiant as the Sun; the soul, in turn, becomes luminous and illuminative. A person may become aware of the radiance of the Archetypal Flames during meditation. This is especially helpful in extending and accentuating the luminosity within the Soul's Medallion.

Talent grace, used day-by-day in original achievements excelling others in the world, comes from the Medallion of the Soul. The Guardian Angel times the creative talents to man. Spiritual power and spiritual gifts come from the Diamond Medallion of Grace around the Higher Self. The Recording Angel, working with the Archetypal Flames, distributes in right timing spiritual power and gifts of the Spirit.

Meditation and the Higher Self

> *To apply oneself to rhythm in meditation, so as to reach an accord with the soul and with the Higher Self, requires an undeviating regularity as to timing and method.*

The Higher Self or Eternal Self is the Instructor of man. The Higher Self is at home in the Spiritual Worlds. To have access to or to reach the Higher Self, one should build a way or a means of communication. Until alignment is made with the Higher Self, the heart, the thought and the soul have individual methods of communication and are often separate in their expressions. Prayer, contemplation and meditation enable the heart, the thought and the soul to become at one, whereby a continuity of love and light may be sustained. A telepathic rapport with the Higher Self is accomplished—and the true voice may be heard interpreting the Spiritual Worlds, the Will of God and His Plan for man.

Through the continued practice of prayer, contemplation and meditation, the atoms of the physical, emotional and mental bodies are brought into alignment with the Higher Self. The great Eternal Laws are interpreted to the disciple, and the Higher Self becomes the logos or interpreter of the Spiritual Worlds.

Meditation should be practiced in an environment of privacy and quiet. The one meditating should be seated, his eyes gently closed. To concentrate upon any one or any thing in meditation is to make the senses the perceiver; therefore, there should be no thought upon any desire or person, as this would interfere with the releasing of tension from the feeling and thinking of the one meditating. It would also interfere with the Higher Self's sending onto the magnetic field of the brain the pure telepathies and the reflections of the Archetypes from the Spiritual Worlds.

Meditation is for the purpose of bringing the various atoms in all bodies into alignment and of building and sustaining a power-body of light. During meditation there is an unseen activity which is later manifested in the outer consciousness in perfect timing to the need. This appears in a form of telepathy as guidance and instruction. In higher stages of evolvement, the Archetypes may be read during the practice of meditation.

Lesser metaphysics sees the demands of the singular ego as being all. Archetypal metaphysics sees mind as a composite, versatile conceptual manifestor for God and in God.

God as Eternal Spirit is patient beyond man's comprehension. Archetypal thinking reveals to the initiate that a self-concentrated psychical mentality is an impatient nature and therefore alienated from God. Archetypal metaphysics, stemming from Universal order, reveres and observes progressive processes of the laws of cosmic timing, seeing in the prototypal action of man the yet unborn and potential spiritual nature. He sees the spiritual mind to be a divinely organized mind seeking its own best within a theme of extensive homogeneous identification.

Meditative reading on arising, prayer selflessly sounded, and meditation will unite the initiate with all of the twelve spiritual aspects of his mind, that he may make union with the Divine, and thus express the highest possible and most blessed prototypal effort.

Retreat from the challenges of life and resistance to the spiritual interflow between the Divine and the physical produce unrest, discontent, dissatisfaction, and works of unreality. Acceptance of Divinity elevates, inspires, cheers and frees.

One should take hold of the twelve mind-mansions of the soul with a right wholeness or holiness, claiming all of his spiritual inheritance with right-centered beliefs, knowing and affirming. He should contemplate the *all* in himself to understand the all of God in the self he calls the *I*.

Until the disciple has made perfect alignment with his Higher Self, his spiritual Earth-Teacher imparts instruction to him and interprets each phase in the disciple's evolvement, aiding him in the resolving of his karma. Alignment with the Higher Self enables the chela to enter into the highest of all instructions. Through the Great Immortals and the powers of Mediation, he makes alignment with the Higher Worlds and the World Teacher, Jesus. When he has made alignment with the Jesus One, he aligns himself with the Archetypal Worlds and the Light of the Christ—and receives of the Eternals.

The Higher Self contains in its hierarchy-will nature a mind sheath supporting the Archetypal Design for that which is to be. Through the mind-willing of the hierarchy nature, the Archetypal Design will not change or be changed from that design. *It will be:* This is the Divine Constant. The True Self is at home in God; not one particle willed by the Creator can be moved. It will happen. It is going to happen.

He who finds his own hierarchy nature cannot fail. In each incarnation, his creation, when responding to the Archetypal Plan in the hierarchy nature, assures that he will succeed, for God is present as the Omnipresence in all his affairs, personal and spiritual.

Is it not written in your law, I said, Ye are Gods?
—St. John 10:34

4

THE SEVEN GENESIS-LEVELS

Do not expect anyone to be or to produce what he is not; always realize that no one can be more than he desires to be. Atom evolvement determines where one is in the Archetypal Light and what level of Genesis he is functioning within.

Before this earth is concluded there will have been seven stages of Genesis on earth: (1) Tribal or Nomadic-Genesis, (2) Family or Human-Genesis, (3) Self-Genesis, (4) Cosmos-Genesis, (5) Pro-Genesis, (6) All-Genesis, (7) One-Genesis. Many in the world today are yet in Nomadic-Genesis and Human-Genesis. However, in this present spiral of action, all men are stirring and seeking to rise into higher levels of evolvement.

True disciples are either approaching Self-Genesis or are in the latter part of Self-Genesis. Only a minor number are in the last part of Self-Genesis—and only 33 in the world are in Cosmos-Genesis.

Jesus of Nazareth is the Divine Prototype or Blueprint which all men are seeking to fulfill in this earth or eternity. *"Jesus saith unto him, I am the way, the truth and the life; no man cometh unto the Father, but by me." (St. John 14:6)*

When men are in the Genesis known as *Pro-Genesis*, they shall be "like" Jesus. *"Beloved, now are we the sons of God, and it doth not yet appear what we shall be: but we know that, when he shall appear, we shall be like him; for we shall see him as he is. And every man that hath this hope in him purifieth himself, even as he is pure." (1 John 3:2,3)*

The first three stages of Genesis cover a period of billions of years. They are designed to accomplish three things:

I. Physical form—a body for action
II. Personality—a body for experience
III. Individuality—a body for thought

In the beginning of man's atom evolvement on earth, his state of expression was *Emblience* or Tribal. As he evolved into human environment and associations, atom evolvement became that of *Niblience* or Family-Genesis. Man now strives to bring forth that of Self-Genesis through *Niscience*. In this, men grow more acutely aware of one another as of individuality. This may be seen in the world, in society, in the family life, and in the personal experience.

In this Earth or cosmic eternity, all men are in some state of Genesis. The balance within the system of evolution from sentience to Niscience is on the average experienced as a form of immersion or absorption; for example, Nomadic or Tribal-Genesis is now blending into Human or Family-Genesis; and, on a higher scale, Human-Genesis is blending into Self-Genesis. From this comes the conflict in which instinct and self-will in the Nomadic and Human-Genesis scale strive to strike a harmonious note in the emotions of man. And in the higher stage of Human-Genesis into Self-Genesis, the note seeking freedom and expression is that the lesser will be balanced into higher thought and mind.

Atom evolvement is the evolution and bringing forth of *Luminosity Atoms* to Light. The first stage of atom evolvement is the fiery sentient laggard atoms in the state of Emblience or Nomadic-Genesis. In this stage, men evolve through instinct and fear.

The second stage of atom evolvement, Niblience or Human-Genesis, is the bringing forth of the laggard atoms into the alternating or duality Archetypal releasings through cell-like pictures. In this comes the pole between the positive and negative, love and hate, and the belief in death as death. The fear within Nomadic-Genesis comes forth as hate within Human-Genesis. And in the increase of Human-Genesis within the state of Niblience, the cunning mind comes forth.

The third stage of atom evolvement occurs when the emotional-body atoms begin their alignment to the etheric-body atoms and bring forth the triplicity cells or Archetypal pictures. In this manner, man's rational mind comes forth, his will becomes individualized, his thoughts take a plan in coherence. This is the beginning of Self-Genesis.

In the fourth stage of atom evolvement, the previous triplicity cells begin a multiplicity cell work. This is made possible through the etheric-body atoms blending with the emotional-body atoms. In this is genius creation brought forth. This is the latter stage of Self-Genesis and the preparation for discipleship. When this has occurred, the disciple becomes aware of a spiritual affinity within a broader range of experience. He touches the Master's Aura or Garment and prepares for his work as proto-disciple in which he will in Cosmos-Genesis become co-aligned to the Disciple Overdwelling Atoms working with the Cosmic Jesus, and therefore linked to the Cosmos Christ.

In Cosmos-Genesis, man becomes one of four Prototypes: John the Beloved, Peter, James or the Jesus Man. He will work in all of Cosmos-Genesis through Love within the higher mind. In this, he prepares for the phase of Genesis called *Pro-Genesis*, in which man shall begin his creation as of mind through thought. His mental body, having brought forth its totality in atoms, shall recapitulate the Love within Cosmos-Genesis into the works as promised by Jesus of Nazareth: "*Verily, verily, I say unto you, He that believeth on me, the works that I do shall he do also; and greater works than these shall he do.*" *(St. John 14:12)*

> *In each Moving-Archetypal period of 10,000 years, each form of Genesis is recapitulated onto a higher spiral in evolution and refinement. At the peak of each Genesis' ripeness, all strong prototypes when perfected are, through the plan in evolution, extracted onto higher spirals.*

THE INDIVIDUALISTIC ATOM

> *The First Heaven is a holy circumference of rainbow-like Light encircling the Earth. After the First-Heaven or Soul-Level Initiations, the initiate who has the atom evolvement moves into higher degrees of Light; he sustains his night-flight experience for longer and longer intervals until he knows Heaven face to face.*

Nomadic-Genesis is for the purpose of giving man the coordinated use of the senses and a perfected physical form as an instrument for physical action, in preparation for the ages to come. Through wars between tribes, he begins the evolvement of an *Individualistic Atom* in his physical body. This Individualistic Atom in his physical body works with the microscopic form-image within the seed-atom of the heart and gives him a consciousness pertaining to his physical body. It makes man aware of his physical body and what it may do for him. It makes him aware of the strength of his physical body. It makes him aware of all things having a form. It also makes it possible for the physical form to be reproduced from life to life.

In Tribal or Nomadic-Genesis, man's ideal is preservation of the tribe. Therefore, he expresses his thinking, sentiently, through the sense-body of the tribe. In this, his own senses, former etheric instruments dulled by the impact of gravity of earth, are perfected and thus more acute. In seeking to preserve the tribe, he develops his senses and a certain instinctual intelligence.

Human-Genesis is for the purpose of man's producing a personality. The personality is not self-created; it is a product of the ages. The personality of each man is produced by the thread of both reverent and irreverent tribal and family associations through countless ages. Even though the personality changes from life to life, the various memories brought forth from previous lives determine the field of personality experience and physical conduct in each life. It is the memory of physical action that determines one's physical conduct in each life. Through this he expresses the instinctual will. It is the memory of his personality associations that determines how great or how small the field of his personality experience may be in each life. Through this he expresses the senses. Until man acquires reverence, the instinctual will and his senses determine his field of personality experience and sphere of physical conduct.

Certain karmic associations in family ties, competitive aggressiveness in the outer world, and wars between nations enable man to produce an Individualistic Atom in his emotional body. This Individualistic Atom in his emotional body enables him to express a certain degree of intelligence in which the emotions are predominant. In Human-Genesis, man's ideal is preservation for the family. Thus, he expresses his thinking emotionally through the emotions and feelings of the family. Family and blood relationships are repeatedly experienced until perfection is brought forth as to the ideal in the human association. The Individualistic Atom in the emotional body enables man to stabilize his emotions and to express his personality through the family so that he might receive reverence. It is the purpose of the whole family association that he acquire reverence. When he has accomplished reverence, he becomes an outstanding personality and therefore helpful to humanity at large.

Self-Genesis is divided into two parts. The first part of Self-Genesis is for the purpose of man producing individuality; the second part of Self-Genesis is for the purpose of man producing selflessness. Whereas personality is produced from tribal and family association, individuality is produced from associating with persons outside the tribe or family. The unfamiliar and non-blood relatings work to produce individuality for man.

In the first part of Self-Genesis, an Individualistic Atom begins to manifest in man's mental body, and he develops a positive and strong individuality. He becomes detached from the former family encasement and seeks to blend with society rather than family. In this, he loses the intimate love he had formerly for his family association. His action moves into the thought world, preponderantly intellect; his emotions are muted and inarticulate. The intellect being predominant, he becomes more critical in his attitude toward love and its demands upon him. He expresses wrong expectations toward love, expecting to be loved in spite of his critical mind and selfish actions. In his desire for individuality, he fails to fulfill the expectations of others of love. He develops strong guilt feelings due to certain unfulfilled family relationships, and a tremendous strain is placed upon the mental body, endangering his mental health.

In some rare instances, genius is manifested in the first part of Self-Genesis. In this, one has the danger of personalized egotism.

In one of the early phases of Self-Genesis, man's ideal is preservation of his right to think and act as an individual, and his right to explore every degree of life. From this phase of Self-Genesis has come the scientific age.

In certain earlier phases of Self-Genesis, there is a period in which man's temperament swings to the side of the pendulum completely opposite to reverence. Thus, he expresses irreverence for the proven things of the past. He attacks and becomes antagonistic to the decaying ideals centered in family or tribe. He seeks, intellectually, to preserve the ideal in self-discovery and experience. Irreverence and intellect become the critical mind. He despises historical inheritance, religious and educational. He fortifies himself only through self-opinion. And he would lay to waste everything which has formerly produced the cultures of mankind, its growth and evolvement. Out of this come revolutions in governments, separateness in families, upheavals in religions, and unreasoning prejudices are accentuated against races, creeds and religions. Fortunately for society at large he is in the minority.

As man gradually evolves in Self-Genesis, this irreverence begins to be rationalized through certain fundamental laws, becoming apparent to the logic within his thought world. The violent hidden or open rebellion against tradition and its inheritance is tempered into a higher phase of reasoning, accompanied by a degree of reverence for the worthwhile attributes of man as to feeling and thinking. Out of this come the higher phases of Self-Genesis, or the man in the world who would preserve an ideal which presents to man his right to feel reverently, to love selflessly, and his right to worship and find his God through direct relating of the heart and mind.

In the second part of Self-Genesis, he begins to teach and impart instruction, showing what the ideal man can be so that others in the world may, in their relating to God, seek to become the ideal man.

In the last part of Self-Genesis, one strives for impersonality rather than personality. His sense-faculties become aspects of the soul. He uses the higher degrees of emotions and thought. And rather than the expression of individuality, he strives for selfless serving and dedication so that in the coming Genesis, called *Cosmos-Genesis*, he may become a Being and express through unobstructed Mediation in alignment to the Christ Mind.

Jesus of Nazareth was free of personality and its expression because His perfection needed not experience in this earth. The personality is solely for man's experience through the periods known as Tribal or Nomadic-Genesis, Family or Human-Genesis and the earlier stages of Self-Genesis. When man has fulfilled Self-Genesis, he shall have perfected individuality, or that is, his particularizing course of action for this eternity. And personality and individuality shall cease to be the chief actors on the stage of the drama of life. Man will have absorbed tribal instincts, personality and individuality into his Higher Self and he shall become a Being in earth.

The Individualistic Atom is the clue between experience and consciousness. Were it not for man's Individualistic Atom in each of the first three Geneses, all atoms within his bodies would remain latent, dormant or laggard, and he would be devoid of any self-directed action. He would be wholly influenced by mass impulse and, thus, the initial purpose of this eternity would be defeated, as it is necessary for man, in the first phases of his evolvement, to become individualized so that he might eventually fulfill a unique kind of hierarchy work in earth.

A person may be out of alignment to his Individualistic Atom either temporarily or for many lives. If so, he is caught in a vacuum-like state. For example, when one in Nomadic-Genesis is sealed away from the Individualistic Atom in his physical body, he becomes a weakling in the eyes of the tribe. In Human-Genesis, when one is out of alignment to the Individualistic Atom in his emotional body, he becomes an indecisive personality, unable to love in any degree. He is yet non-self-reliant, therefore parasitically inclined, and is dependent upon family and family opinion. When faced with vital issues, such persons fail, as their structure is no stronger than that which they relate themselves to in their family association. If one in Self-Genesis is out of alignment to the Individualistic Atom in his mental body, he is yet unable to organize his thought process or think with clarity or justice.

When a person is said to have a magnetic personality, he is expressing the Individualistic Atom of his emotional body. When a person is thought of as a strong personality, fortified by wisdom and truth, he is in alignment to the Individualistic Atom of his mental body.

If I so choose, I can enter into the discipline given of God, and gain the Kingdom of Heaven.

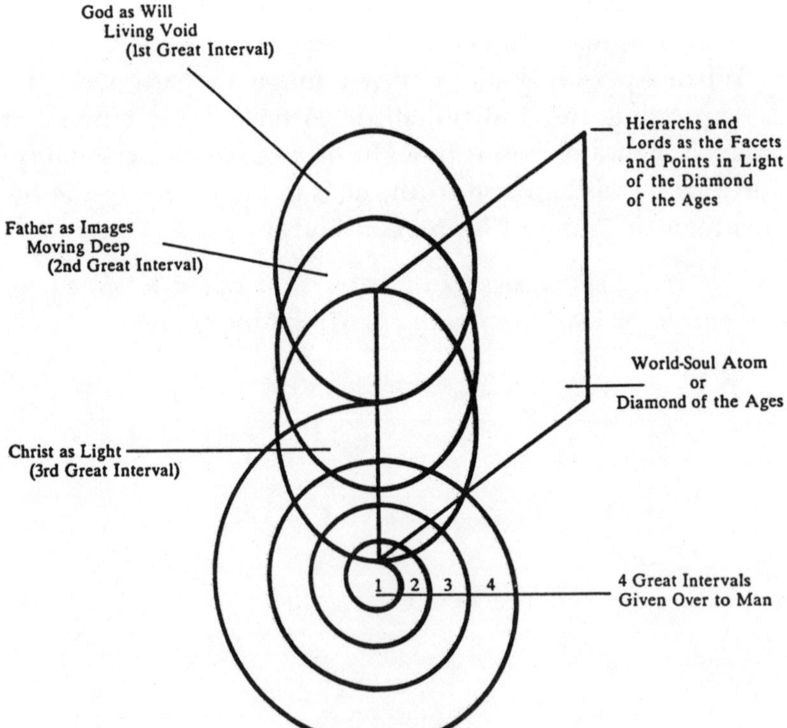

THE ETERNAL'S WORK IN THIS COSMIC ETERNITY (EARTH) THROUGH THE WORLD-SOUL ATOM OR DIAMOND OF THE AGES

5

THE WORLD-SOUL ATOM

There are false jewels and there are jewels fair. May I wear the pure jewels of the Spirit for my breastplate of love.

The World-Soul Atom or Diamond of the Ages

The Central Atom of the earth is the World-Soul Atom. Until the coming of Jesus, the Father and Hierarchy worked through the World-Soul Atom; since the Resurrection of Jesus, the Christ-Spirit has taken command of the World-Soul Atom. "But Jesus answered them, My Father worketh hitherto, and I work." (St. John 5:17)

The World-Soul Atom, working for the Father, externalizes the Moving Deep onto man. In this, as a diamond, it shapes, forms and extends, and is the means of leveling Space into expansion and expansion into Time. As the Indestructible Atom between the brow is the axis of man's soul-light, the World-Soul Atom is the axis of the Eternals for man.

The Father differentiates the Images to come forth in this Earth or cosmic eternity through the World-Soul Atom. The World-Soul Atom uses the rhythmic expansions correlating to all Images, timed to the procession of their coming forth. The World-Soul Atom brings the Images forth that man, in his evolving within the four Great Intervals, may receive and experience these Images. The physical Sun of this cosmic eternity makes it possible for the Invisible Sun of the Christ to send forth the Images of the Father into Light as Archetypes. These Archetypes are activated by the Christ into action in Earth.

The Hierarchs and the Lords as the facets and the points in Light of the Diamond of the Ages, aiding in the work of Creation with the Father, ray Their Light through the World-Soul Atom onto man. The Diamond of the Ages or World-Soul Atom receives its geometrical outline and means of movement by the Hierarch streams of Light into its proportion as of Light, Tone and Sound onto the impulses of man. These streams of Light, falling into the Spheres of Light, are in accord with the World of Tone and make up the Music of the Spheres which, in turn, come to man as sound. The Indestructible Atom between the brow of man, being a lesser replica of the World-Soul Atom, works as a diamond facet to transmute the power of Light and the forces of sound; and, through the shafts of Light, sends forth the individualization as of Archetone and sound unto man.

The World-Soul Atom is the differentiator of the energies in this cosmic eternity. In Nomadic-Genesis and Human-Genesis, the Will of God works through the World-Soul Atom in man as the will to live. In Self-Genesis, the Will of God works through the World-Soul Atom to bring forth the Higher Self and the higher will. In Cosmos-Genesis, the Will of God works through the disciple as Love. In Pro-Genesis, the Will of God works through the disciple as Manifestation.

The action of the Ultimates is determined by the World-Soul Atom. The Ultimates work as karmic, rhythmic interludes. In Earth, they are the Will of God as Law and externalize themselves as the Laws of Sowing and Reaping, Retribution, Cause and Effect, Consequence or Karma.

FROM NOMADIC LIFE-WAVES TO FAMILY

The planetary rays, working through the World-Soul Atom, work with the over-group rhythms to bring forth particularized prototypes for man. This occurs in all nomadic life-waves. In the latter part of each nomadic life-wave, the stronger prototypes are brought forth. From these, the planets begin an extraction by activating the laggard atoms within the prototypes.

When the laggard atom in its fiery sentient state has released itself from the Over-Will within the nomadic life-wave, the first Individualistic Atom is brought forth in the strong prototype. The strong prototype from Nomadic-Genesis then comes forth in the repeated births or embodiments preparing for the trial in Human-Genesis, in which he individualizes and expresses the traits and characteristics that define the trend or inclination expressing in a family. In this, he also begins his first work of the lesser will, through which he evolves for the family, perfecting that which is individualized as family, as in all Human-Genesis each family has an over-atom working through the blood and lesser etheric body. This creates the strong attachments and bonds of "blood being thicker than water."

The planetary rays overshadow the family, working to release the fiery sentient atoms and bring forth the purer forms as typified in family or group. In this day and time of individualization as of the Christ, strong or possessive attachments by blood carry heavy karmic penalties, as man now seeks to release Self-Genesis and to free the Individualistic Atom and the higher will.

Conflict and separateness within the family group will be replaced by honoring individuality. Family, in Self-Genesis, shall express individualization and respect onto one another. Love will have risen to a higher spiral, in which detachment shall attract into families purer egos, who, as catalyst associates, shall create within the rise of the spiral of evolutionary cultures.

Man's Work with Genesis

In Nomadic-Genesis or Emblience, men in tribes are overshadowed by the species impulse or Over-Will which works through the World-Soul Atom. Their evolution is through the fiery sentient atoms as sentience. These work on the physical body as of form.

In Human-Genesis or Niblience, men in families are overshadowed by the Archangels and Angelic helps who work through the World-Soul Atom for the differentiation of prototypes. In the beginning of Human-Genesis are the laggard, fiery-sentient-atoms ones, who in self-will work through the lesser etheric body, the lesser emotional body, the five senses, and the lesser planetary energies. In the latter part of Human-Genesis, with the freeing of the laggard, fiery sentient atoms, the higher-emotional-body atoms evolve and begin their blending with the seed-atom of the heart that they may combine with the higher-etheric-body atoms and give Hierarchy the means of differentiation as to strong prototypes being brought forth. There are few today who have evolved beyond the latter stages of Human-Genesis, and many are still in the stage of Emblience or Nomadic-Genesis.

In the latter part of Human-Genesis, the emotional body is being prepared for Self-Genesis. In the average man the emotional body is still in an ovoid, rubberball-like state, which, through resistance, frictions and tensions within, creates a self-contraction, and, through the aid of Hierarchy, seeks to bring forth the atoms for future individualization as prototype-genius. In this, the emotional body is created into a formlike structure. It is through the perfecting of this emotional form that man shall begin to build the coming mental body. His work in Self-Genesis is directed to this goal.

In Self-Genesis, man begins his first knowledge of karma as self-responsibility. *The overshadowing of the World-Soul Atom is replaced by the Cosmos Christ, Who works to free man from the instinctual will. "My Father worketh hitherto, and I work." (St. John 5:17)* Man begins to face himself, and thereby brings forth strong individualization and genius.

When one is rated eccentric or in any manner different, this is but the beginning of the dissolving of egotism, which, by difference, singles him out from others, and is but the beginning of a long, hard trial for true individuality as of Niscience.

When one creates something uniquely individualistic or brings forth a value hitherto unexperienced by man, he may be said to be beginning Self-Genesis.

The overshadowing of Niscience as intuition aids in the creating of Self-Genesis. Within the latter part of Self-Genesis, the future Cosmos-Genesis disciple will begin his work for the acquiring of Niscience, or the memory of the Eternals, that he may bring forth the atoms to unveil and reveal the mysteries within God. As the five senses make the five fingers of the hand to reach, to test and to prove the physical, the five atoms within the higher emotional body, when in alignment with the two pinnacle emotional-body atoms, make it possible to reach into Niscience and to recall former experience in other cosmic eternity systems.

John, the beloved Disciple of Jesus, is the pure, perfected Prototype that man shall strive for in the latter part of Self-Genesis and in the beginning of Cosmos-Genesis. The proto-disciple shall work within Love and the Love-Ray, of which Jesus in earth gave to man the keynote to discipleship.

In Cosmos-Genesis, man comes under the Cosmos Christ

and works with the Father, his work being on world-karma, the aiding of the resolving of karma, the bringing forth of the laggard atoms in men into the Light of the Christ; and in the working with the Christ, to aid men in the absorption between the World-Soul Atom and the Christ-Spirit.

The Cosmos-Genesis disciple will come forth after man has brought forth the emotional-body Love-atoms and the control of his mental body. This will occur when man has perfected a fluidic matrix and brought forth the latent mental atom. Man shall think, knowing the origin of thought, determining the measure by which he thinks, and the result of what he thinks. When this has come, man will have acquired the *Omniscient Cell* within the brain and shall use the power of Niscience, or knowing beyond knowledge.

In Pro-Genesis, man will fulfill the perfected Jesus-Man Prototype. Having brought forth the mental body, and having acquired the pure Love-Niscience and the Love-Prototype of John the Beloved, man shall be blended with the Cosmic Jesus, the Cosmos Christ and the Father, fulfilling the saying of Jesus, "*At that day ye shall know that I am in my Father, and ye in me, and I in you.*" *(St. John 14:20)* This shall be the Day of Manifestation and the beginning of the fulfillment of man's inheritance of the Earth.

My lamp and wick are ready, clean. I await the touch of the flame kindled by my intense desiring.

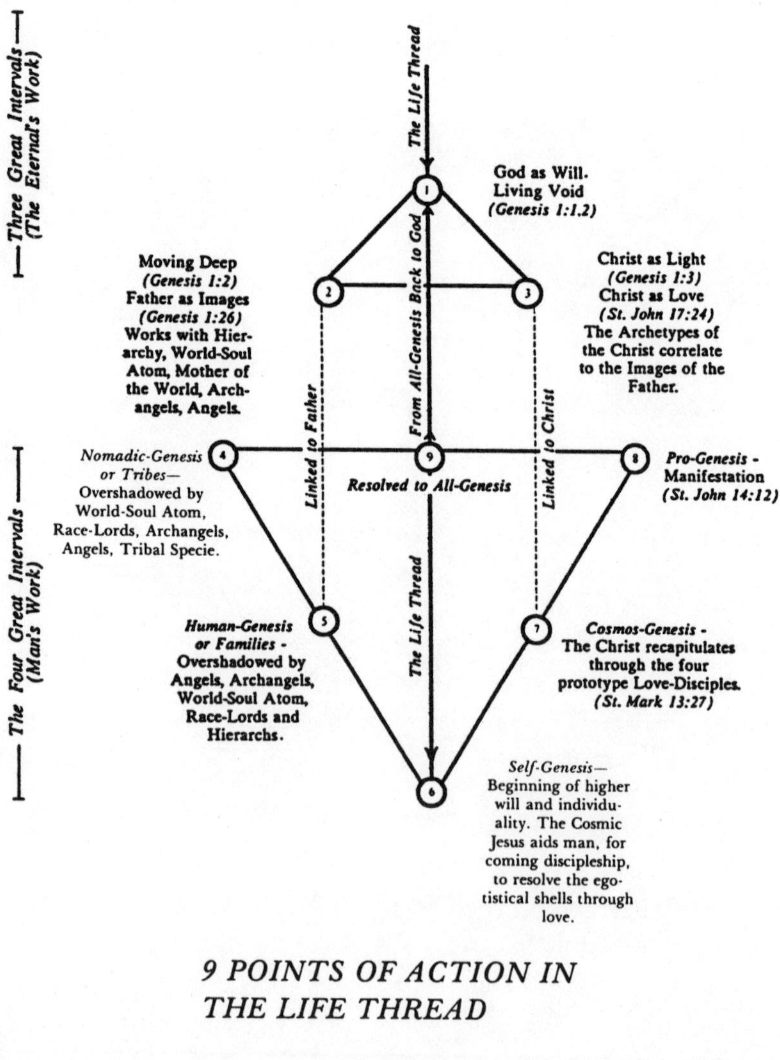

9 POINTS OF ACTION IN THE LIFE THREAD

6

ARCS OF THE AGES

There is a method, a formula, a rhythm. Let me find the true, for I would build temples rather than castles.

Ascending Arcs

The new Genesis Age will live and express a new Ethic under Christ.

Men work in Self-Genesis to free the dynamics of the mind through the channel of love.

Herewith is given the history of man's ancient origin and of the akasic thread of survival through countless lives and expressions, beginning at the Tribal and slowly evolving through the ages to produce the present personality of this period. Each of the ascending arcs, as portrayed in the following Genesis Charts, represents a Great Age. For example, the Tribes portion of Nomadic-Genesis represents 6 Great Ages; the Races portion of Nomadic-Genesis represents 6 Great Ages. Therefore, Nomadic-Genesis, Human-Genesis and Self-Genesis total 36 Great Ages, covering billions of years through which men have evolved in this earth.

On the left or rising side of each arc (see Charts I, II, III), the six symbols represent thousands of lives of one prototype (man) and portray his struggle to achieve the crest of the arc and therefore leadership. As men move upward on the struggle, effort or talent side of the rise in the arc, they express countless personalities, seeking always to reach the peak of the arc. However, many by-pass this leadership peak, absorbing but only a fragment of that offered to them out of the Will of God and His Plan for man. Such men are the laggard ones in earth.

The right or descending side of each arc is the absorption side of the arc, and portrays the ages of tempering or modifying the will. Thus, the right or descending side of the arc seals into man the mercy attribute in Nomadic-Genesis, the compassion in Human-Genesis, and the ethic in Self-Genesis.

It is the work of the disciple to unravel the akasic thread and weave it into comprehension, that he may understand the various stages of evolvement and of the seemingly unending processes making up the personality of man. The disciple who would aid, heal and also instruct should look with reverence and wisdom into the akasic record of this eternity so as to retain his balance and thus be sustained in unlimited Mediation.

Tribes

A. The rising side of each arc in Nomadic-Genesis is struggle. Man is but one cell in a tribe nucleus. He struggles through many lives to become a Tribal Leader. In this he produces the instinctual will through the use of the senses, and has a primitive, etheric alignment to the elemental forces, the Tribal Dweller and the World-Soul Atom. He worships the forces of Nature.

B. After countless lives he begins to bring forth the Individualistic Atom of his physical body, thus becom-

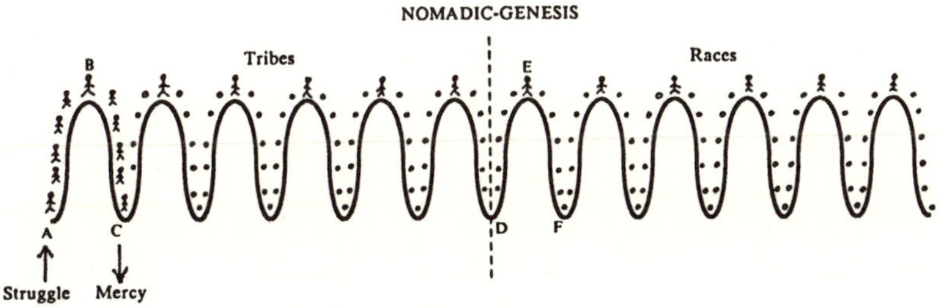

Chart I: Nomadic-Genesis

ing either a Tribal Leader or benefiting from those who lead him.

C. On the descent side of the arc, his lives contain lessons of Mercy so that the physical-power urge gained in his previous rise and achievement in leadership may be tempered and balanced.

RACES

D. In the Races portion of Nomadic-Genesis, one struggles to achieve Patriarch Leadership.

E. He becomes a Patriarch Leader or achieves some authority.

F. In his decline or descent from leadership, he works to become a peacemaker between Tribes and Races.

Re-embodiment. In Nomadic-Genesis, man re-embodies in life-waves or masses of humanity. From life to life he alternates his sex from male to female, male to female.

Karma. The Law of Karma is less individualized and works through the tribe and the race. In this, men work through a common conscience of the race rather than individual conscience. Debts of karma are balanced through cataclysms, famines, floods, wars.

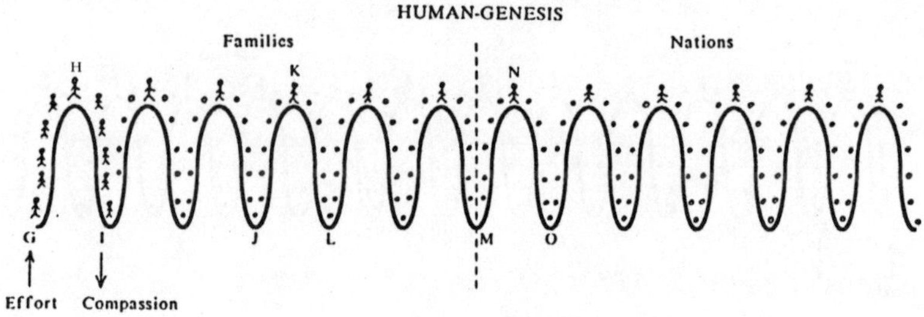

Chart II: Human-Genesis

FAMILIES

G. On the left or ascent side of the arc, man exerts the effort to free himself from the former Nomadic-Genesis expression and the compelling influence of the World-Soul Atom. He establishes communities, villages, and becomes encased in a Family-Atom; his relationship to his offspring begins to be more refined. Through repeated trials in family association, he begins to produce personality as a vehicle of experience.

H. On the peak or crest of the arc, he makes alignment to the Guardian Angel and Recording Angel of his Family-Atom, thus becoming a Family Patriarch. From this comes the means through which he intuits the needs of his family. And he becomes directly responsible for the moral purity of his ties in blood.

I. In the descent from the peak of the arc, he inherits the conscience of the Family-Atom. Through this conscience aspect, which works through the Recording Angel overdwelling the Family-Atom, he begins to incorporate a compassion for the suffering of other men; and over the ages, a humanitarian attitude toward his fellow man begins to imprint itself upon him.

J. In the ascent side of this arc, he starts a strenuous and deliberate effort to purify the generation or procreative stream. He begins to bring forth an Individualistic Atom within his emotional body.

K. When he reaches this peak or crest of the arc, he revives old etheric powers he had in previous bodies. In some instances he may become prophetically aligned to that which is beneficial for his family or families in his environment. He builds a tabernacle, and establishes sanitary laws and protective family laws for ages to come.

L. He becomes a priest within a temple and establishes certain ritual through which men worship.

NATIONS

M. He establishes religions, education, cities, governments, nations. In the latter part of Human-Genesis, men seek God through personal relating and interpretation. They personalize God and worship Him as a personal Deity.

N. On the peak of the arc, he may become a religious leader, educator, statesman; his authority is directed toward the good of humankind.

O. He is contested by wars from other nations. In this he works out former race and family karma and strives to balance the racial impulses in the world. Until now he has used the instinctual will, but in the descent side of the arc he has the danger of using the cunning mind and cunning will, thus endangering that which has been built in previous ages. In this phase of Human-Genesis, he has the opportunity to choose between the use of pure ideals for mankind as a whole or to use the cunning mind to preserve his personalized relatings for the inclusive interest of family.

Re-embodiment. In the first part of Human-Genesis, one still alternates his sex from male to female, but he begins to re-embody from one life to another in a slightly quickened timing. In the latter part of Human-Genesis, a woman may remain in a feminine body or a man may remain in a masculine body for many successive lives until a certain quality is perfected.

Karma. Karma works in a more sensitized manner in Human-Genesis. The Law of Attraction is accelerated and man begins to meet his karma on more individual levels. Without deviation, the pendulum of retribution attracts to each one that which he has sown. Thus, the opportunity is given to each one to resolve his own karma.

SOCIETY

P. On the left or ascent side of the arc, man works with talents brought through grace from many former lives. He strives, through individuality, to fulfill perfection in talent and thus become genius. He relies solely upon his egotism and trusts only his intellect. He worships his own intellect, as his intellect is his God.

Q. In rising to the crest of the arc, he seeks to bring forth the Individualistic Atom in the mental body. In this he expresses strong individuality and, in some rare instances, genius. Because he is encased in an egotistical shell and is yet to bring forth pure love, his danger is the basking in self-acclaim and self-love.

R. On the descent side of the arc, his major point of action is the war between logic and a hidden hunger for love. He works to resolve his egotistical shell and to balance thought with love so that he may enter into the second or discipleship phase of Self-Genesis. He is, in reality, seeking to find the ethics in each thing, that he might perceive his own ethic and therefore prepare himself for the latter stages of Self-Genesis in which he will express pure ethics. He begins to see God as The Intelligence and The Plan.

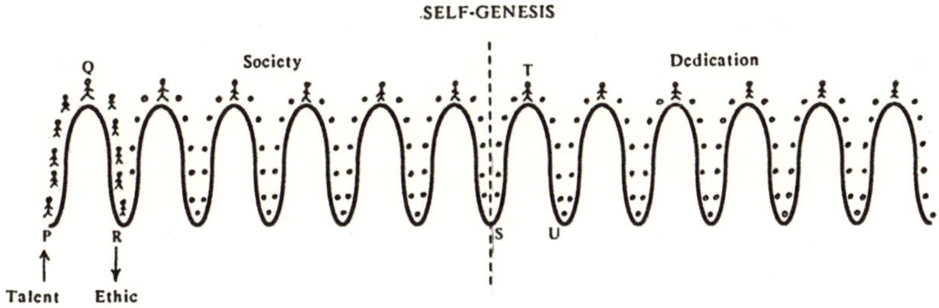

Chart III: Self-Genesis

DEDICATION

S. On the ascent side of the arc, one seeks to channel his serving so that he might selflessly work as a world-server to overcome world-karma. In his seeking to serve, he is confronted with his unresolved karma of Nomadic-Genesis, Human-Genesis and the first phases of Self-Genesis. From this come frequent Dweller experiences, presenting to him certain degrees of opposition or karmic contest in his rise to a spiritual life. These may be presented from race, family, nation, society or from individuality levels of former action and experience.

T. On the crest of the arc, he achieves discipleship and aligns himself to grace and the use of spiritual gifts gathered out of this eternity experience. He gains his ethic through which he may serve.

U. On the descent side of the arc, he becomes at one with the experiences of mankind. He works selflessly to dedicate to the Eternals and to align himself to the Christ Mind, that he may heal, serve, instruct and teach through the Archetypes and Will in God. He finds God as The One.

Re-embodiment. In Self-Genesis, re-embodiment is accelerated and man moves more quickly from life to life. In the latter part of Self-Genesis, the dedicated disciple who is uniquely fitted for certain tasks may, previous to birth, choose his future sex, environment and vocation.

Karma. In all phases of Self-Genesis, karma moves at a very rapid pace and, for the disciple, there is instantaneous reproving as to feeling, thought and action. In the latter part of Self-Genesis, his karmic labors and responsibilities are heavy in that he must incorporate certain karma from former Genesis levels. And rather than becoming indifferent to others, he becomes impersonal as well as loving, blending in whatever environment he may be placed. Karma is clearly perceived by him. His goal is to live a life of dedication and selfless serving.

The Jesus Ethic is the Door to the Path—
and He who holds the Door is that Example
and Archetypal Blueprint for all to follow.

GENESIS TRAITS: TRIBAL-GENESIS, FAMILY-GENESIS AND SELF-GENESIS

No one may enter the needle-eye door to the Kingdom of God save he has "clean hands and a pure heart." (Psalm 24:4)

The following outline is presented to the disciple to identify certain root-traits and tendencies now being expressed by men in the world so that in this work with others he may determine what degree of Genesis is being expressed by them. The disciple should especially note the negation words as related to Tribal-Genesis and the lesser degrees of Human-Genesis. Should he find these root-traits outpictured in his own life or the lives of others, he should remember that these negative and decaying traits, when not resolved, are yet saturated with certain atavistic ethers which man once used in combatting the world on a level little known in our time. For example, in the early stages of Tribal-Genesis, such traits as hatred, fear and rivalry were necessary protectives measures to produce certain strength capacities needed to master the forces of Nature and the animal world. However, when these natural traits of Tribal-Genesis are retained by men who are beginning to develop the cunning will and cunning mind in Human-Genesis, rather than natural traits they become fierce traits which are dangerous to mankind.

Tribal-Genesis and Human-Genesis had both reached a decadent stage when Jesus came into the world. It is when these decaying periods are experienced in the earth that great Saviour action is made possible. In such a time came Jesus as Saviour and Messiah to give men *"the way, the truth and the life." (St. John 14:6)*

All men in the earth in Human-Genesis and the earlier stages of Self-Genesis are colored by some of these primitive traits. Some of these traits create a very real combat area in evolvement; others move in subtle or insinuative manners; and others are shadow-like—but all or each of these may stand between the disciple and his evolvement. He should never underestimate the danger of these hidden primitive compulsions within. It is the work of the disciple to resolve the root-traits which would possess him in certain intervals or crisis periods of his life. Until he has resolved these ancient etheric compulsions within, he will be their servant rather than their master. If the disciple indulges even a remnant of hatred, it will stand between him and the Light. The work of the disciple in the earlier stages of Self-Genesis is to face fearlessly any lingering or remaining negation root-traits of Tribal-Genesis and Human-Genesis and to recognize them as to what they are. Before the disciple may evolve to higher labors in discipleship, he must determine the "what I am" so that he may become the "I am."

GENESIS TRAITS

Tribal-Genesis	*Lesser degrees Human-Genesis*	*Higher degrees Human-Genesis*	*Lesser degrees Self-Genesis*	*Higher degrees Self-Genesis*
Courage	Bravery	Fearlessness	Daring	Renunciation. "*Whosoever will lose his life for my sake shall find it.*" St. Matthew 16:25
Crude	Reformed	Refined	Sensitive	Soul Health. "*Be ye therefore perfect, even as your Father which is in heaven is perfect.*" St. Matthew 5:48
Cupidity	Avariciousness	Acquisitiveness	Contemptuousness	Stewardship. Respect of others' belongings. "*For what is a man profited, if he shall gain the whole world, and lose his own soul?*" St. Matthew 16:26
Curse	Oath	Vow	Promise	Covenant. "*If any man will come after me, let him deny himself, and take up his cross, and follow me.*" St. Matthew 16:24
Fear	Cunning-Will	Cunning-Mind	Manipulation	Death to the Primitive Will. "*Father . . . not as I will, but as thou wilt.*" St. Matthew 26:39
Fervor	Zeal	Ambition	Initiative	World-serving. Humanitarian.
Fortitude	Endurance	Stamina	Mastery	Overcoming. "*Be of good cheer; I have overcome the world.*" St. John 16:33
Greed	Jealousy	Covetousness	Speculation	Acceptance. "*For he that hath, to him shall be given.*" Mark 4:25

GENESIS TRAITS

Tribal-Genesis	*Lesser degrees Human-Genesis*	*Higher degrees Human-Genesis*	*Lesser degrees Self-Genesis*	*Higher degrees Self-Genesis*
Hatred	Animosity	Prejudice	Critical-Mind	Consideration. *"Therefore all things whatsoever ye would that men should do to you, do ye even so to them."* St. Matthew 7:12
Honor	Character	Integrity	Standard	Ethic. *"Inasmuch as ye have done it unto one of the least of these my brethren, ye have done it unto me."* St. Matthew 25:40
Indigent	Dependent	Competence	Independence	Worthiness. *"By their fruits ye shall know them."* St. Matthew 7:20
Instinct	Apprehension	Foresight	Intuition	Seeing clear and true. *"If therefore thine eye be single, thy whole body shall be full of light."* St. Matthew 6:22
Isolation	Hypocrisy	Bigotry	Unteachable-ness	Illumination. The perceiving heart and the discerning mind.
Labor	Duty	Responsibility	Self-sustaining	Grace works. *"Labour not for the meat which perisheth, but for that meat which endureth unto everlasting life."* St. John 6:27
Lethargy	Slothfulness	Inhibited	Indifference	Gratitude. Impartial appreciativeness.
Lust	Lasciviousness	Sensuousness	Lack of self-control	Purity. *"Blessed are the pure in heart."* St. Matthew 5:8
Malice	Persecution	Gossip	Slander	Logos. Chaste speaking.

GENESIS TRAITS

Tribal-Genesis	*Lesser degrees Human-Genesis*	*Higher degrees Human-Genesis*	*Lesser degrees Self-Genesis*	*Higher degrees Self-Genesis*
Pride of tribe or race	Pride of blood or stock	Pride of possessions	Pride of intellect	Humility. *"The Father that dwelleth in me, he doeth the works."* St. John 14:10
Primitive, possessive love, (children and wives are as chattel)	Personal, possessive, jealous love, (husband or wife feels "ownership" of mate)	Love of possessions	Self-love	Selflessness. *"A new commandment I give unto you, That ye love one another."* St. John 13:34
Rage	Anger	Antagonism	Indignation	Self-control. *"Bless them that curse you."* St. Matthew 5:44
Revenge	Vindictive	Retaliation	Self-righteous actions	Forgiveness. *"If ye forgive not men their trespasses, neither will your Father forgive your trespasses."* St. Matthew 6:15
Rivalry	Opponents	Competition	Strategy	Victory through initiation. *"Agree with thine adversary quickly."* St. Matthew 5:25
Treacherous	Timidity	Self-doubt	Egotism	Self-denial. Willing sacrifice for others.
Tribe or race as self	Family as self	Personality as self	Self-importance	Soul-equanimity. Selflessness.
Trust	Blind-faith	Belief	Proof	Prescience. Niscience or knowing.
Vulgarity	Irreverence	Reverence	Imbalanced judgments	Pure perceptions. *"Judge not, that ye be not judged."* St. Matthew 7:1
Will to survive	Will to live	Will to experience	Will to express	Serving as a Mediator. Will to serve.

7

THE AGE OF SELF-GENESIS

I came not to send peace, but a sword.
—St. Matthew 10:34

Self-Genesis

The individualistic person in Lesser Self-Genesis is concerned with his own evolvement; he is engrossed with his own self-interest. In Higher Self-Genesis, he recognizes the right of every man to become an identity and to relate himself to the Cause of his being, or God.

The world, turning on its axis, is now accelerating in its core with a higher vibrancy caused by the Christ Spirit, which took command of this eternity system with the Resurrection of Jesus. From this has come the Self-Genesis Age which has as its first urgency and vital concern the preservation of self or individuality. The first concern for more highly-evolved souls who come to contribute to the Self-Genesis Age with their soul-powers is to find Light and to make union with God.

In the Age of Self-Genesis, men will pride themselves on thinking for themselves; there will be great emphasis placed upon the power of thought, sometimes called the power of mind. From such mental concepts will proceed many swift lessons and casualties, for he who expands his thinking for self and self alone brings to birth objects and things out of timing. Metaphysics, or the organizing of the thought, is a necessity in the Self-Genesis Age, and will enable man to come closer to the imaging power or creating power given of God; but until man unites his heart, his mind and his soul with God, his thinking will produce painful and partial results.

During the last century, self-interest has been the central theme in governments and religions. It is apparent that consideration for others has been complacently set aside and self-interest has been uppermost. One of the casualties of the earlier part of the Self-Genesis Age is that men are more interested in themselves than in their fellow man. In the higher stages of Self-Genesis, men will express the mercy aspect toward one another.

The personal guru will become obsolete in the Age of Self-Genesis. However, there is to be a finer association between Teacher and disciple; this will be a co-atom telepathy. The true disciple will work to become co-atom to his Teacher, and thus make alignment to a source of mediation little-known in this time.

As one evolves into Self-Genesis, the symbology which has been so important to him in Tribal and Lesser Family-Genesis begins to re-enter his emotions and thoughts. The etheric symbologies dormant in the lower part of *Quelle* or *Center Q* at the base of the skull are activated. At the present time, these atavistic symbols are intensely interesting to psychiatry. These symbols, when analyzed, often show primitive repressions and traits. To a psychiatrist with a discerning mind, these symbols also prove that man has a conscience.

In this Self-Genesis time, man of the earth is given the

opportunity to unite his ego-gained faculties with the Most-High-Soul powers. This was made possible through Christ Jesus who came to overcome the world that man may know the world as it is in all aspects of human and soul life. Through interrelatedness, perspective is gained and the new man in Christ is born.

Men are still very limited in God-Realization, especially so in Lesser Self-Genesis. However, in the eternal Time and tuning fork of God, it is an Ultimate that men who are created of Light must return to the Source of Light, which is God.

Man illumined is in Higher Self-Genesis. He is a trans-energizer of the gross energies allotted by the senses.

If one is in Lower Self-Genesis, he magnifies his ego. If he is in Higher Self-Genesis, he spiritualizes his intellect, seeing God as the One in All. If he has reached Cosmos-Genesis, he is a Cosmos-Disciple, and he has established a likeness of John the Beloved, the closest disciple to Jesus.

The Bible—the sacred Book given to man in its Old Testament aspects for the Family-Genesis phase of life, and New Testament for a Self-Genesis expression—is watched over by great Recording Angels who see to it that this sacred book may be, chameleon-like, suitable to any situation in the life of man. Until men have reached the completion of Self-Genesis, this Book will remain sacred to men, supporting all things pertaining to their spiritual progress and their physical moral conduct.

In the Old and New Testaments may be found references to the upward rise in the prototypal climb. The twelve sons of Jacob represent the Tribal-Genesis prototypes. The twelve prophets of the Old Testament of the Bible represent the Family-Genesis prototypes. The twelve disciples around Jesus represent the Self-Genesis prototypes. Each one of these twelve in Tribal-Genesis, Family-Genesis and Self-Genesis is a prototypal reflection of Hierarchy. Men on earth today are seeking to blend and perfect these twelve prototypal resemblances and expressions within themselves so that they may incorporate and blend all of these into one, and thus become the perfect and pure prototype, even as Jesus was perfect. When this has occurred, men shall be "like Him" (1 John 3:2)—for Jesus is the ultimate Prototype which men may fulfill on earth.

Jesus is the Door for the Higher Self-Genesis soul who is seeking to make union with the World of God and thereby experience God-Realization.

LESSER SELF-GENESIS

For I am come to set a man at variance against his father, and the daughter against her mother, and the daughter in law against her mother in law.

—St. Matthew 10:35

As one enters the first phases of Lesser Self-Genesis, his individuality is more accentuated, and he often loses touch with his higher intuition and his faith. To him, his individuality and self-importance are foremost. Being devoid of the naive ideals he formerly had, he feels and thinks through a materialistic image—he thinks upon what is due him, rather than what he can do for others. His sexual life is one of experiment and exploring. He expects others to give to him. He knows little of sacrifice. He is uncomfortable when the soul is mentioned, as he thinks that belief in a soul places him into a dependent meekness unsuitable to him. In his conversation, he is sarcastic, curt, caustic, often twisting words to gain his ends. He sees other men through the eye of cynicism. He looks upon them as "suckers." And if they are more successful than he, he is envious of their "luck." He has forgotten there is spiritual power, and he is allergic to spiritual persons.

The Lesser Self-Genesis person, lacking reverence and faith, produces a materialistic age. His body, his emotions and his soul are subjected to a life of force, hate and pressure. When the initiations of Lesser Self-Genesis are concluded, one enters into a certain austerity affecting the sexual life. He learns to love again and to see the sexual act as a sacred, reverent action. In his appetites and habits, he practices self-denial. He gradually learns to release possessions, persons and self-will.

Lesser Self-Genesis men bent on having their own way at any cost, expressing themselves as being the only one, began a wreckage-way of life, assaulting the very morality which had supported Human-Genesis orthodoxy and traditional ways of life. The majority of men in this age and time are expressing Lesser Self-Genesis. All are eventually, however, to enter into the Higher Self-Genesis way.

To live in Lesser Self-Genesis, devoid of a full and active conscience, is an unhappy and bitter life—a most lonely and desolate life. Regardless of how much culture one has attained, or regardless of his refinement in association, one without conscience is always seeking something more. His true hunger is for constancy in friendship and association.

When egotism is transformed into humility, one acknowledges that God governs the Universe, and that God alone holds the answer to his desolateness. He holds up his conscience to God, and God revives it. His soul-powers become active; he begins to move freely between the lesser and the higher promptings of his conscience. He begins to understand when a "no" means "no." He also begins to understand when a "yes, yes, yes" means "yes."

In Lesser Self-Genesis, choice and will have more free play than in any other degree of Genesis. Therefore, Lesser

Self-Genesis is a painful state; it is, however, a necessary stage in evolvement. God set man free with choice and with a certain degree of will. God gave the conscience to be the watchtower and the watchman over man's will. When one understands that self-will is nothing unless it is immersed in God's Will, then his Angel comes forth and he begins to receive holy guidance—which works directly with his conscience. His conscience becomes prescient conscience, and he has reached the state of Higher Self-Genesis.

In Lesser Self-Genesis, one begins his work of self-discovery. In this, there is an accentuation of self-centeredness. Jesus, the most selfless being ever to live in the world, taught men to deny themselves and to take up their cross and follow Him. In Higher Self-Genesis, one learns to recognize the difference between conceit and selflessness. Humility and self-denial give him the means of freeing his true being given of God. The Higher Self-Genesis person channels his works, emotions and intellect into a purposeful and spiritual life.

In the present age of birth to Lesser Self-Genesis, the mental and emotional neuroses and the unpredictable amoral acts are caused by extreme internal pressures; the initiatory trials through emotions and mentality are accelerated. The world conscience, the family-atom conscience, and the individual conscience press heavily upon those who have been laggard in responding to the spiritual Laws ruling the Universe.

A Lesser Self-Genesis sickness is arrogance expressed as stepping upon the feelings of others. Failing to see that one can offend in his own actions is a Lesser Self-Genesis disease.

Every true disciple of the inner and higher life should seek to temper his hypersensitivity by recognizing that he himself may have fault in his own nature.

One may recognize with ease the Lesser Self-Genesis person, as he is represented wherever there is psychological or psychiatric stress. The Lesser Self-Genesis person is self-engrossed; he is personally centered in expectations of a successful life rather than in spiritual being. The Higher Self-Genesis person is selflessly centered in the universal life. The predominance of self-seeking persons in this age generates a selfish society bent upon license and personal preference.

The weight of a collective Lesser Self-Genesis evolvement is placed upon the moral and spiritual shoulders of advanced souls. Self-selected placement in a shadowed family-atom doubles the life burden of the more highly evolved ego; sometimes this produces potential sainthood, enabling him to lift by a supernatural love the souls sharing life with him.

The Lesser Self-Genesis person, as a rule, must yet learn the ethical use of the will. His compulsion to explore and to learn for himself, if defeated, often culminates in some form of will-sickness. When one has misused the will in former lives, and is born to a Self-Genesis time, he is sometimes frustrated, bitter.

The Lesser Self-Genesis man depends more on what he experiences and what he thinks he knows. He is contemptuous of tender approaches to his soul, and he is blind to the laws of righteousness existing in the world. He is often cynical—and distrustful of others. He is prone to blame or to put the burden of his mistakes on others, as he is deaf to the voice of his conscience.

The Lesser Self-Genesis person has much of the Lucifer angels in his nature. He is prideful; he is in love with himself. He often procrastinates in his better nature of hospitality and of love. He is either partially loving, or he is lustfully loving, or possessively loving.

The Lesser Self-Genesis man thinks that conscience is the

product of various disciplines by traditional bodies—religious, parental, and collectively national on various levels. He does not know that conscience is a spiritual asset to which he has access, and by which he may guide his ship and come safely home.

As men become more egotistical in Lesser Self-Genesis, they receive ego-shattering lessons that return them to their conscience. On awakening to the conscience, one dies to egotism. Very often there are many tears of self-reproach and states of melancholy; there is much grieving, and the revealing of dimensions of selfishness in one's nature. This is a period of dying to the *egotistical shell.*

In the Lesser Self-Genesis Age, the most prevalent sickness is that of mental depression and mental imbalance. One cause of mental suffering is the shadowing of the egotistical shell against the soul's light. The egotistical shell has been built from self-centered acts in this life and in many past lives.

A Lesser Self-Genesis person who is disobedient to the voice of conscience expresses self-deception, dishonesty, unreliability, amorality, agnosticism, egotism and frenzied inconstancies.

In Lesser Self-Genesis, man is concerned solely with consciousness; self-awareness predominates over conscience. The conscience of a Lesser Self-Genesis man must speak to him through subtle or indirect ways. One who deliberately turns away from responding to his conscience—because he feels that he is submitting his will to traditional concepts—experiences dreams which are heavily laden with reminders from his conscience. Through dreams his Guardian Angel sees that he is ever reminded of wrongdoing.

The mental and emotional suffering in Lesser Self-Genesis is more intense than in any other Genesis.

Many egos preparing to enter Lesser Self-Genesis carry heavy marital soul-debts from past lives. These debts are presently being expiated by some through accepting the burden of a loveless marriage or through assuming the burden of divorcement and separation.

When men begin to evolve into Lesser Self-Genesis, marriage stands forth paramountly on trial. The Lesser Self-Genesis Age is the center of chaotic pressures, producing unstable associations and relationships on all levels of human behavior. Morality, fidelity, honor, protection, responsibility—these appear to be empty phrases to egos born into certain Family-Atoms.

In the Lesser Self-Genesis Age, accumulative marital soul-debts are the cause of persons divorcing and remarrying one or more times, for such persons are unable to establish the complete Family-Atom relationship. These persons, needing companionship, enter into each marriage with a desire for happiness. However, until all accumulative marital soul-debts are erased from the vibratory hum of the Soul's Medallion, they find it impossible to alter the drifting current of their emotional hungers.

The Lesser Self-Genesis person married to a person of Family-Genesis level of evolvement does not understand unison of ideas relating to marriage, children and the home. Should the Lesser Self-Genesis mate be the husband, the results would be an aversion to the purchase of a home or permanent residence, a neglect of marital demonstrativeness, and an indifference concerning the increase of children; in some cases, conceiving too many children without thought of the welfare or the health of the mate. There is irresponsibility as to finances and the ordinary necessities of the family.

The Lesser Self-Genesis husband, being more concerned with the drive of his ego, is not willing to sacrifice to protect his family, but accepts the sacrifice of his family and children as his just due.

When conscience is awakened in a Lesser Self-Genesis person, it is often difficult for him to transmit to his children his new set of values. He finds it difficult to communicate with his children as a responsible parent, or to communicate with them as a person. This is a tragic situation, in which not only the parent suffers but also the family and the children.

The strong, hostile currents in a Family-Atom in the Lesser Self-Genesis Age produce an unstable society. In the age of supposed reasonableness, vast numbers of Lesser Self-Genesis egos demand indulgent panaceas for their indigent ways. A resisting mentality is an irritating cell in the flesh of society and of nations, and as such continues until some form of higher ethic is established. The spiritual scientist trained through initiatory formulas in the inner planes will work on the outer planes to produce a balance. It is the work of the initiate to strive to activate the living principles and ethics established by Jesus.

> *Even as those who did sit in the Upper Chamber changed their Arc of Genesis, so hast thou changed thy Arc of Genesis. And thou hast the way on which thou dost rise, for from the Christ is the arc of the rise. And from timing cometh the Genesis which overcometh the long, wandering way. The accelerated timing of the Christ bringeth man into that place where he may resolve the Genesis memory of the old.*

HIGHER SELF-GENESIS

> *Jesus is the first and Supreme Avatar for the world. All who have reached Higher Self-Genesis consciously enter the door to instruction through Jesus.*

In the latter part of Self-Genesis, one comes closer to the Christ-impulse, and thus becomes penetrable to the honeycomb filigree action in Mediation. Thereafter, he begins his true expression and is at home in the physical world as well as in the Spiritual World.

In the second part of Self-Genesis, one desires to teach and impart instruction, showing what the ideal man can be —so that others in the world, in their relating to God, may seek to become the *ideal man*.

In Higher Self-Genesis, the conflict and separateness within the family group are replaced by honoring and respecting the individuality of each member of the family. Love is expressed on a higher spiral. Such families consist of purer egos who, as catalyst associates, create within the rising spiral of evolutionary cultures.

The Higher Self-Genesis initiate expresses purity in sex, self-denial. He has absolute faith in the Will of God. He loves his fellow man and those near him who share his love. He encourages men and inspires them through his healing words of love. He thinks with an illumined mind, seeing always God as the Creator, as the One.

When men have attained the Higher Level of Self-Genesis, there shall be introduced into the world a perfect rhythm of daily living. This will include five working hours, five recreational hours, five creative hours. From this, man will achieve the highest ethic as to life and its purpose. Needless to say, this perfect rhythm for the physical life is impossible for the average person in the present society. However, all persons who would utilize their lives to the uppermost should pray to attain a perfect rhythm expressing the creative life.

When a person achieves Higher Self-Genesis, he is constantly aware of the presence of his Angels and their ministering helps. And he begins to align himself with the Kingdoms of Angels and their work with the world.

The Higher Self-Genesis person seeks to channel his spiritual gifts so that he may selflessly work as a world-server to overcome world-karma. In his seeking to serve, he is confronted with his unresolved karma of Tribal-Genesis, Family-Genesis and Lesser Self-Genesis. From this come frequent *Dweller* experiences, presenting to him certain degrees of opposition or karmic contest in his rise to a spiritual life. These challenges may be presented from race, family, nation, society, religion—or from his personal soul-debts of the past.

A person in Higher Self-Genesis achieves soul-powers and frees his higher levels of grace. He begins to use the spiritual gifts gathered from aeons of living in this eternity, and he gains the ethic through which he may serve.

The Higher Self-Genesis man becomes at one with the experiences of mankind. He works selflessly and dedicates to the Eternals, that he may think through the Christ Mind, and thus heal, serve, instruct and teach through the Greater Archetypes and Will of God.

Higher Self-Genesis or Soul and Spiritual initiatory dream symbols appear when one has overcome the lesser levels of Genesis evolvement. Soul-and-spiritual symbols in dream dramas portray heavenly attributes and mediative associations, keeping the balance between the physical and spiritual worlds. In soul-and-spiritual symbology, one experiences his dreams as instruction rather than apprehension. The one dreaming beholds the beauty of the Higher Worlds and remembers the manner, voices and intonations of his instructors of the night. He moves into wave lengths of revelation, observing without fear the visions of the dark. He is given ritualistic power through certain dream-code understanding, that he may be protected from the dark. His spiritual powers in the day expand into creative action of the day.

In Higher Self-Genesis, one comes into communion with the spiritual aspects of his dream symbologies, and he becomes familiar with the trustworthy symbology of mediative action through dreams as well as through daytime action. He begins to communicate with the *etheric braille* of the Spiritual Worlds. From this he extends his knowledge of the world of the soul; he learns of the action of his soul, of the meaning of his soul, and particularly of the meaning of God. God's life in him is symbolized through the inspiration he gives to men in the world.

In Higher Self-Genesis, symbols are received from the grace level of the soul's record, from the record of the Higher Self, and also from the true records of the Archetypal Worlds. The soul-logos symbology speaks to the highly evolved person through contemplation, meditation, and through higher dream levels.

When Higher Self-Genesis comes to full expression, symbology will instruct and also reveal. With spiritual sensitivity one may eventually attain one of the greater symbolic or spiritual arts—the power of the parable. Such persons, having used the symbolic arts through many lives of spiritual sensitivity, can induce upon the thought power of an attentive listener the actual and living image within the spoken word. The Lord Jesus used this power of image-parable. His words will never die because they were directly empowered by the master images within the Greater Archetypes dwelling within the Third Heaven or the Kingdom of God. *"Heaven and earth shall pass away, but my words shall not pass away." (St. Matthew 24:35)*

The Higher Self-Genesis man will eventually unite his thoughts with the Christ Spirit, rather than with the World-Soul Atom—and he will become an Apostle for God.

Pentecost is the fulfillment of the Ascension of Jesus. *"And I, if I be lifted up from the earth, will draw all men unto me." (St. John 12:32)* The Higher Self and the soul directing the ego repetitive patterns began to express through mankind through a Self-Genesis state of consciousness.

With the lifting power of Jesus in His Ascension, souls having advanced consciousness through past-life incarnations entered into Higher Self-Genesis expressions; spiritual powers and gifts surfaced, projected from their souls. The Esse poured down upon them the Passing Power, that they might keep open the processes of soul-awareness of God.

In the latter stage of Self-Genesis and individualization, man strives to dissolve all remaining karmic, prototype shells, and to prepare a fluidic matrix for the use of the higher will and the higher mind. During the latter part of Self-Genesis, man is overshadowed by the twelve perfected disciple prototypes, who were the disciples of Jesus in Earth, and who now, as co-atoms to the atom body of the Cosmic Jesus, dwell in the Realm of Light. In this overshadowing, man begins the trials onto discipleship, preparing him for the coming Cosmos-Genesis.

The loins, being the center for the genes, generation and genesis, are the focal point for the four Genesis Dweller Trials: (1) Tribal-Genesis, (2) Family-Genesis, (3) Self-Genesis, and (4) the satanic dweller before attaining Higher Self-Genesis and eventually Cosmos-Genesis. When a disciple is prepared to rise to a higher degree of evolvement, he undergoes a major initiatory trial within the area of his loins. When a disciple dedicates to live more reverently and to unveil the purified flame of the heart so that the soul may speak, the initiatory process begins. He may experience more than one state of former Genesis action in various manners: (1) through dream experience, (2) through health, (3) through associations in karma, such as persons in his life, (4) or he may be subjected to certain temptations presented to him through amoral and unethical persons.

The Immaculate Conception in Mary, officiated and manifested by Gabriel and the Holy Spirit, signifies the birth of the Higher Self-Genesis in the human race. The human race is reaching toward the birth of the Omniscient

Cell in the Gleaming Brain. When this occurs in the Earth, scientific and spiritual creators will be cognizant of Galaxy-functioning as affecting life in this Solar System.

In the present humanity chaos, all men are striving to raise their degree of Genesis. The United States of America is the first nation to strive for Self-Genesis. The Self-Genesis expression in the United States of America is little understood by the Tribal-Genesis and Family-Genesis nations. In time, a balance will be set, and harmony among the humanities will return.

The strong egos in the United States are divided into two levels of Self-Genesis: Lesser Self-Genesis and Higher Self-Genesis. The Lesser Self-Genesis person is inclined to be egotistical and self-sufficient, and is generally disliked by other peoples in the world. The Higher Self-Genesis ego, as yet a rarity, is charitable, ethical. To the Higher Self-Genesis individual will be given the task of upholding the Archetype of the United States of America.

The United States of America is a virginal crucible for a Self-Genesis humanity. Thus, this nation has taken upon itself much of the world-karma so that the Self-Genesis Age may be born in the world. The penetration of the United States into Korea was to prepare the Eastern and Asiatic soul for the transition into a higher degree of Genesis.

The Self-Genesis expression in the United States of America also will manifest an altogether different rhythm between the sexes. The man of Self-Genesis will become more temperate; the woman more self-sufficient. There eventually will come to the United States of America a higher form of marriage ideal. From this a great nation will emerge.

The United States is an extrovert nation; India is an introvert nation. The Self-Genesis people in the United States of America hold the key to the polarity for the new continents of the earth.

Higher Self-Genesis ideas will come to the masses in the

Aquarian Age and will begin to manifest outwardly in the year of 2568. Higher Self-Genesis ideas are presently being expressed by the initiates under Seraphim. Their light, known as the Phoenix fire of Heaven, quickens the mentality impulses of everyone attracted to them. They are the Initiators or *Maha-Matras* of their age—known in Heaven as "the twinkling-of-the-eye Initiators." Blessed is he who has been called to instruction under the Phoenix fire.

> *Those who now enter the Self-Genesis phase of evolvement are beginning to aid the Christ in the assimilation and absorption of the World-Soul Atom. From this will come the heaviest labors in evolvement.*

ANCESTRAL GRACE AND HEALTH

The spiritual healing arts in Lesser Self-Genesis require an enlargement in prayer, faith and understanding. Lesser-Genesis egos denying the ancestral survival supports may be exposed to many health trials through the body, the emotions and the mind.

Ancestral survival grace fails to completely support or sustain one who feels himself to be beyond the influence of the Family-Atom. In all intimate and rightful associations, God has set a restoring principle through which one may conjoin something of his bodily and physical needs. Through associated inter-working blood relationships, sympathy and empathy from blood relatives play an important part in all healing. When sympathy and the love of a mother for a child are denied, this checks and nullifies the interflow of ancestral struggle for survival and renewal. Should a mother fail to be sympathetic to her child, the angelic healing helps are thwarted. Where sympathy is lacking, the ancestral instincts fail to fully play their part in the survival of the sick.

Moving over the Lesser Self-Genesis experience with love, respect, and reverence for one's origins and source of life unites one more firmly with the Angel of his body—the Luminosity Angel. To seal away by deliberation the flowing chain of ancestral survival helps is to shut away millions of years of upward survival instincts. Through Lesser Self-Genesis callousness to unconsciously hate or deny one's origins is to mitigate or lessen the will to live.

Lesser Self-Genesis is self-deceptive in that one thinks himself to be beyond interior supporting human influences in things of the body, emotions and the mind.

Belief in the Luminosity-Angel body-helps for bodily health will reinforce the will to live. Mantrams designed to unite with the Luminosity Angel give the power to merge the body impulses with the strength of the best of inherent traits, both physical and spiritual.

O Luminosity Angel of my body,
If it is the Will, give unto me
The grace strengths of my ancestral traits,
That I may arise and do
The Will of Him who healeth all.

THE HOLY GHOST

Receive ye the Holy Ghost.

—St. John 20:22

Through the feet, man is a magnetic willing instrument. In Nomadic or Tribal-Genesis, Holy Ghost guidance-brain impulses thrust upward the subtle, serpentine, Kundalini astral fire through the feet into the nervous and muscular systems of tribal man. The sacral biological impulses compel men to propagate and to beget tribal offspring expressing *sentient thinking. Tribal-Genesis thinking seeks to protect the tribe rather than the individual person.*

In Human-Genesis or Family-Genesis, Holy Ghost, working with Elohim Hierarchy, centers His power upon the navel-plexus guidance brain of Family-Genesis men, causing them to become specific zodiacal thinking prototypes such as Aries, Taurus, Gemini, Cancer, etc. *The thinking of the Family-Genesis man is directed to society, community, family property, and prestige of family.*

In Lesser Self-Genesis, Holy Ghost plays His tones of power upon the solar-plexus guidance brain of man. *Lesser Self-Genesis thinking pertains to thinking one's self to be all-important, thinking of one's own interests before the interests of other men. The Lesser Self-Genesis person is inquisitive and self-centered, producing selfish works. Hardened hearts produce unteachable minds.*

In the Higher Self-Genesis person, Holy Ghost centers His work upon the Sacred Atom of the heart. The love emotions of the heart are then channeled into all of the feelings, thoughts and organic centers of man. Holy Ghost works that the heart of man may produce the twelve apostolic forms of love, and thereby enable man to attain Higher Self-Genesis thinking. *Such persons will have access—in essence only—to the guidance-brain preservative and protective instinctual-thinking telepathies and instinctual tribal memories, and thus be free from the hostile quelle impulses of the past.*

In Higher Self-Genesis, the love body is united with the higher mind. The thinking of evolved men moves through the pure, etheric, cosmic mind matrix of souls, where separations of minds are impossible—blending all minds in the flowing Christ-vortices of Spirit. *Higher Self-Genesis thoughts see all men or souls united under God—akin as sons of God.*

> *All phases of higher intuition are activated by the Holy Ghost, that man may become a perfect seer for God.*

LAW

O how love I thy law! it is my meditation all the day.

—*Psalm 119:97*

The Lesser Self-Genesis person comes into the world psychically pressured by his karma of past lives, by the karma of races, of families, and by the karma of the world. Under such pressure, his grace or his karma determines his rightful or erring attitudes and actions while living in the world.

Those less evolved in the individualization processes of Lesser Self-Genesis are inclined to use the judgment aspect of their minds rather than the perceptive side of their minds and wills. These may be seen in the present day's acts of chaos on the stage of life as the radicals, the liberals, the dissenters, the malcontent, the violent, and the revolutionary.

Law for man and Universal Law are as one with God. Differentiations of law and the effects of equation functioning impartially through the processes of God's Law upon the person and upon the world can best be understood through a relaxed, perceptive and unifying mind.

The spiritual aspirant understands and perceives justice and law as the two necessary functions mercifully balancing, leveling, blending, maintaining the Cosmos, the Cosmic and the personal with wisdom. He keeps his own life processes fully engaged within the equalized Divine-averaging Law which never ceases. He can do this only by becoming an accepting balancer within, by being a relaxed navigator of his own evolvement. He sees the judging tendency in his own nature to be ever in need of mitigation and softening—the need for charity, patience, forbearance. Through the upholding of love, through right discrimination rather than harsh judgment, his perception becomes keen, pure; and as a faultless, sensitized instrument he leaves judgment to Him who judges all.

Universal Law is set into this world to protect man, to maintain his forward-going works without deviation, to sustain the Divine averages, that man go not beyond the preordained vision and Plan as given of God. Trust in Law as protector, as upholder of the Universe, produces peace, calm, and ultimately perfection as envisioned for each soul in the earth.

All greater Saints and Avatars knew Law, and unto the very last breath left judgment to the Maker and Upholder of the Universe. To use wise perception with prayer in all matters—physical, personal and spiritual—is to live within the Law. He who lives within the Law will abide in a perfect perception as creator for God.

> *In Higher Self-Genesis, the love-body is united with the higher mind.*

GENESIS AND RELIGIOSITY

> *But the hour cometh, and now is, when the true worshippers shall worship the Father in spirit and in truth: for the Father seeketh such to worship him. God is a Spirit: and they that worship him must worship him in spirit and in truth.*
>
> —*St. John 4:23,24*

One should open himself to the religiosity in *all* Scriptures, that he may experience the worship-impulses contained in all Scriptures.

Every 10,000 years in the world, men produce a new Scripture. This Scripture becomes the soul-vehicle for the masses and initiatory vehicle for the individual.

In the beginning of this 10,000-year Moving Archetype, men of the earth in a minor post-Atlantean epoch produced the *Bhagavad Gita*. In this period, men are still under the influences of the Hebrew establishment of the *Holy Bible*.

This Bible, containing the Old Testament, the New Testament and the Book of Revelation, will continue to polarize worship, religious and soul-practices for a period of 3,000 years. In this process, men will polarize the Christ in their natures and extend this knowledge into the Scripture for the next Moving Archetype, which will formulate its own Bible or Scripture.

In each Moving Archetype, men enter into Initiatory Trials for four states of Genesis. That which gave the Scripture for the present Moving Archetype was the Jehovah impulse working through the ancient Hebrews.

The Old Testament, which supports all Scriptures, began with Tribal-Genesis and Family-Genesis Initiation, whereby the Laws of God were established through Moses in the Ten Commandments.

The New Testament as established through Jesus gave to men Self-Genesis, both low and high, preparing them for the fourth stage of Initiation into Cosmos-Genesis. The direction for Cosmos-Genesis Initiation is contained in the Book of Revelation in the Bible.

Men of the earth today are being exposed to the Christ Consciousness. All must eventually come to express Higher Self-Genesis and Cosmos-Genesis.

Religiosity provides interflow within Scripture, whereby one may be initiated at the same time in various degrees of Genesis-levels of consciousness. Those who fail to have knowing teachers regarding Scripture experience these crossing energies of religiosity in one's nature stemming from race, society and nation that create separateness and tumults. A true Living Teacher having religiosity opens his student to the interconnectedness in religion, in worship.

> *Exalt the Lord our God, and worship at his holy hill; for the Lord our God is holy.*
>
> *—Psalm 99:9*

HIGHER SELF-GENESIS TEACHERS

> *Go ye therefore, and teach all nations, baptizing them in the name of the Father, and of the Son, and of the Holy Ghost: Teaching them to observe all things whatsoever I have commanded you: and, lo, I am with you alway, even unto the end of the world. Amen.*
>
> —*St. Matthew 28:19,20*

When a student's motivation for instruction is fixed upon ego-advancement, this reveals to the Teacher what sort of disciple he is working with. When a student's motivation is fixed upon a *cause*, the Teacher finds a student who has entered Higher Self-Genesis.

The Higher Self-Genesis disciple is concerned for enlightenment for all of mankind. The Lesser Self-Genesis motivation is concerned for the preservation of the ego and ego-interest.

The most sensitive points in Initiation in the present-day disciple are the crossing over from Lesser Self-Genesis to Higher Self-Genesis. Teachers who come to initiate such ones have attained Higher Self-Genesis. Being the proteges of Omnipresence, these Teachers are few. When a student is engrossed in ego-motivation, he is blind and deaf to the instruction coming from such Teachers.

The Higher Self-Genesis Teacher treads two waters to accomplish his mission: the waters of Maya and the waters of Purusha.

The Higher Self-Genesis Teacher is initiated in profundities little-understood by ego-motivation. In the life of such Teachers, their wealth and enrichment pertain to knowledge, knowledge nourished and inspired by the Cosmos and the Cosmic.

It is rare to find a Teacher who has both Cosmos and Cosmic literacy. When this occurs, there is harmony enacted, which will be eventually reaped by all souls in an eternity system.

SELF-GENESIS INITIATION

The Gifts are seeking to manifest through the pressure of the soul. In this Self-Genesis period, the Gifts are seeking to return so that man may enlarge himself as a soul and come equal with science and its prolific manifestations.

There are forces in charge of our lives. We must unite with them and use them. When we come in contact with them, we are in contact with God.

These forces are *Intelligible*. They are an Intelligible aspect of God. They reveal themselves through events of testing, refinement and sensitivity. They are not possessed or expressed as personal individuality. They serve that which would become more than the individual—expressing the Increase of God.

In the lesser Geneses, one does not unite with these forces. When one enters into Self-Genesis, his mind becomes receptive to the Intelligible portent of forces and powers which are in reality *soul* energized portions of God.

It is automatic for one who has command of these forces to lead other men to the Light. Their volatile energized voltages pull men forward into increasing awareness of the soul.

It is the destiny of all Self-Genesis persons that they be initiated into the world of forces. This world of forces resides within the Most High Devachan or First Heaven.

The trials and initiatory experiences for the Lesser Self-

Genesis person anticipate the higher energized state of the mind, which is the Higher Self-Genesis state.

All Higher Self-Genesis persons lean in the direction of Cosmos-Genesis. To come under the tutelage of a Cosmos-Genesis Disciple is to participate in the axis turn-over of coming splendors through reciprocity.

Thirty-three Cosmos-Disciples exist in this Archetypal period. Their central polarity points attract to them souls who are in the latter state of Self-Genesis. These become the mitigators for the expanding of the Dharma.

The Greater Archetypal Dharma under command of the Christ Spirit is speaking now to the Cosmos-Disciples and to the Telepathic Disciples.

> *When men reach the Age of Cosmos-Genesis, there will be a separation of the sheep and the goats. The goats (or stragglers) will enter into a twilight sleep; the sheep will begin their work of manifestation, and eventually will be like Jesus, having Hierarchy powers.*

Synchronized Discipleship

Jesus, Master of the Masters, Lord under the Lord of All, needs a worship-vortice through which He may work with the masses. He also needs a body or nucleus of specialized souls to work with individually. In all worship bodies where the Masters synchronize their labors uniquely with Jesus, a very wonderful message can be given to the world.

It is needful to worship; it is also needful to work toward higher initiation that one may truly serve in the world. All initiates will remain in the lesser instruction until they understand that Jesus is the Archstone up-

holding the building of spirituality. Since Jesus, the Masters can do a mighty work, and the greater things men are to do are made possible through synchronization and mediation.

Discipleship is an individualized process. No two evolve the same. Yet in this age there is a singular proven formula relative to all. Through the heavier levels in karma and human relationships, each disciple works to become a mirror or picture of Cosmos-Will. In this, he will reflect and discern the evolvement procedures in discipleship. In the coming Niscience and Self-Genesis, men are to be synchronized in varied disciple levels.

The process of synchronization is a consciousness on a Cosmos level, with an awareness or penetrableness into various levels or degrees in Light. When the Self-Genesis disciple correlates to the original twelve co-atom disciples with the Jesus One in the Realm of Light, the disciple receives the means to synchronize Cosmos into Cosmic and Cosmic into Cosmos. *Cosmic* is Earth eternity; *Cosmos* is God Eternal. *Cosmic* is our Earth and the planets that work with it. *Cosmos* consists of all the planetary bodies and galaxies in Space.

The Earth and its planetary system is but one of twelve cosmic eternity systems, all working similarly. In this, our Earth is co-atom to eleven other cosmic eternities which work through their particularizing systems to aid in the Earth's coming forth, as this Earth is the last facet in a system of twelve cosmic eternities working with one another to perfect a minor Cosmos system in the World of God.

The Lord of Lords, the Cosmos Christ, is the Over-Atom or Shepherd overdwelling all twelve cosmic eternity systems. The Jesus One is the Over-Atom overdwelling our cosmic eternity. The Higher Self overdwells the disciple in his various activities and latent atoms yet to come forth.

The eleven other cosmic eternity systems working with our Earth may be compared to the disciple's atoms now active and perfected in Light. Our Earth may be compared to the mental atoms yet to be brought forth within the disciple's body of Light, as it is the work of this Earth and its planetary system to create through thought in the Christ-Mind and to manifest the Intelligible Will of God.

When Self-Genesis in Earth has become a reality and the disciple has evolved atom alignment to his mental atoms and the Higher Self, he will begin to attune himself to the eleven other cosmic eternity light-streams aligned to our Earth, blending through Niscience. In this there can be no separateness. He is then also blended as a co-atom light-stream activity to all meditative disciples in this earth and to all such energies awaiting to perfect him as disciple in Earth. Through this he receives in Cosmos Light the Higher Self-Genesis Archetypal Realities awaiting him through the Christ-Mind.

Disciples now dedicate themselves to the opportunity to serve selflessly, to devote themselves that they preserve the real within all relatings, regardless of where karma has placed them. Human-Genesis now dying to man and self-interest being the most potent, it is the disciple's labor to keep alive the "love one to another" that men not perish in their egotistical selfishness. The peril of this age in the resolving of Human-Genesis into the Self-Genesis level has brought the avalanche of evil into the world. The dedicated disciples in Earth and the great Mediator Beings within this cosmic eternity's sphere are the karmic levelers and agents through which this balance is to be met, leveled and sustained.

Any self-procedures as related to spiritual values in which mediation is the highest activity level, when related to self-interest, are still working through that of Human-Genesis which now functions through the World-Soul Atom as a decadent activity. Thus, to retain personalized aspiration on any level is to invoke immediate karma related to that which is a dying phase of activity. The result is karma on a painful level of self-experience to explode and resolve the veil of egotism concealing the answer and delaying the synchronizing between that of cosmic eternity and Cosmos Eternals.

Let the disciple of now not only learn to synchronize and align himself to Cosmos but let him learn to penetrate and also to discern the occurrences within himself, as he, of course, is aware that he himself is a cosmic eternity in the making and that each thing in this synchronized whole is but an extended area of orbital activity governed by rhythmic periodicity and coordinating cycles.

Using the power of the will as a chief function results in unprepared astral situations. Such persons are the victims of being overcredulous as to their own right to authority or in the use of their wills over others. A truly synchronized person has been initiated by delicate and refined states in the use of the will within the Will of God.

Those who live steadily and firmly as synchronized vehicles within the Jesus Ethic are repelled by the occult life. Their need for devotion in worship and for rationality in the development of their mental potential through right instruction makes them honest workmen, fellow craftsmen with the Lord Jesus.

SYNTHESIS-HARMONY MARRIAGE

Lesser Self-Genesis is wrecking the Family-Atom.
Higher Self-Genesis is saving the Family-Atom.

The magnetic forces of attraction and repelling are perpetually at work between man and woman. Selection of a mate is more than biological. One's choice of a mate is determined by the soul's need to expand its glorious potential into the human spirit. Regardless of one's concept as to marriage, the great powers of the soul and the spirit work continually to make man's outer being equal to his inner being.

No person should judge another person as to his selection of a mate in marriage. One who is spiritually evolved knows marriage to be a necessary state in the physical world, not only that men may produce children for the human spirit, but that man and woman should be true helpmates united and made one against an alien world.

The masculine and feminine polarities in all persons are projections of spirit and soul. These seek to unite, that the higher progenies of creation or greater ideas may be born to the world.

When there is a perfect marriage on the physical plane, the Angels can fulfill their work for the human family through those who truly love in the marriage state.

Collective traditional restraints built into one's thoughts concerning marriage result in foggy interpretation of the relationships in marriage.

Every 7,000 years the polarity impulse changes in the marriage relationship. In the year of 1846, the masculine-polarity emphasis governing marriage began to change to a feminine-polarity emphasis. Women entered into more aggressive attitudes. Family-Genesis relationships came under severe and disturbing internal stimuli.

In the Lesser Self-Genesis Age, all partnerships in marriage or out of marriage are undergoing transition, for something is seeking to come to birth in a mighty polarity shifting. It is the divine intent that marriage shall come into a holy synthesis, and thereby establish a more ideal state of equal strengths and talents, thus producing a union of harmony and love.

In a synthesis harmony in marriage, the bride and the bridegroom think upon one another as a fusing of the two into the one; they look to God to use their complementary attributes—loving emotional, creative mental.

Until men reach Cosmos-Genesis, the synthesis-harmony marriage will be a rarity. It is grace when one marries a similar, and thereby magnifies his own nature through the person he loves. When one is fortified by magnification of his own qualities through his mate, this is an anointed marriage.

No one in the physical world ever escapes from the law of diversity until he has attained the divine union in his inner nature between his soul and his spirit. Through the diversities one is shaped and formed; thus, when one finds his similar in marriage, they share the diverse trials of the world. United against diversity, two persons living in the Higher Degree of Genesis become a solid front, exemplifying character, integrity.

All initiatory marriages are marriages of opposites in temperament. All synthesis-harmony marriages are marriages between those of similar nature.

When one looks into his interior nature, he begins to understand something of the dual and alternate polarities. Equalization of these dual powers within himself is prompted by the soul and the spirit. Until he has accomplished the equalization of the dual powers, marriage will remain an initiatory state. When one has finally interblended the powers of his emotions and thoughts, he will attract a mate of equal nature and equal temperament.

A healthy, reverent marriage love under the Law of God will stimulate those who are married to love others. Such love is a hospitable love which in no way lessens the love toward one another. It is automatic when one truly loves his mate under God, that he will be more loving, more trusting, more loyal, more reverent to others in the world.

Christian hospitality may sometimes be felt by one mate more than the other. A desire to share one's home and the love emanating therein, when in the heart of but one of the mates, places restraint upon the Family-Atom.

Hostility toward other persons is an atavistic reflex unnatural in a state of healthy marriage. A desire to share one's hearth in joyous intervals of association is proof of one's soul health and equanimity.

Lesser Self-Genesis persons are more inclined to divorce than those living on the level of Human or Family-Genesis. A man and woman expressing Lesser Self-Genesis, both being over-individualistic, are less likely to be charitable and patient with one another. Should there be children from a former marriage state, this will increase the confusion and dissension element within the marriage.

Each marriage has its own individual equation subjected to unseen laws of correction and predestination. It is inevitable that a marriage fail if it is founded upon lust, faulty sentiments, immaturity.

A desire for one person to dominate the will of another in any relationship offends the Law of God. In marriage this is more offensive in the Eyes of God when one person in a marriage, knowing himself to be magnetically irresistible to his mate, uses the power to manipulate the soul of the one entrusted to him through marriage.

It is offensive to God and His Law when a marriage partner devotes himself more to his father or mother parent rather than to wife or husband. It is also against the law of perfect partnership for a father or mother to show particular preference for one child in the family environment. The

Family-Atom is offended when either parent would abnormally possess a child.

The sexual life is a natural life between man and woman when founded upon mutual love and trust of one another. It is offensive to God when husband or wife commits adultery in thought through flirtations, or through dwelling in sensuous thought upon a person outside of the marriage.

> *Unto the pure all things are pure: but unto them that are defiled and unbelieving is nothing pure; but even their mind and conscience is defiled.*
>
> —*Titus 1:15*

> *I dedicate to make reverent my procreative acts through chaste habits.*

In the beginning of this eternity, when the Moon was sloughed off from the Earth, the Eternal Sustaining Atoms, encasing the microscopic blueprints of men-to-be in the earth, began the shaping and forming of the four bodies* of man. As the earth cooled, these bodies enlarged, took form, and became more tangible.

When the planet Mars was sent forth into orbit, the Earth shifted its axis and began to cool and to harden. Man experienced the "fall" from the etheric state. The spinal system, the nerves, the blood, the cells and the glands of man ripened; propagation began. To those less evolved, the procreative act became a mystery; from this came the erroneous belief in "original sin."

The birth of the Lord Jesus gives man the example of the highest quality of birth in the world. In time, many will earn the right to be born through pure mothers and pure

*The etheric, physical, emotional and mental bodies.

fathers. Such persons will come under an angelic blessing. They will be chosen to fulfill a high destiny in the world.

When one thinks sacredly of intimate ties of blood, he builds something for the Angels to work through. When one accepts the fact that man is a noble being in the making, he begins to produce noble relationships in all levels of life.

There are two life currents which influence man from life to life: the life-force current and the life-spirit current. The life-force current is the procreation power given of God for reproduction. The life-spirit current is the connecting and everlasting link with God. Should one seek to irreverently use either current, this becomes an anomaly or an abnormal thing.

When the disciple has experienced and absorbed the Family-Atom Initiation, the next phase of his evolvement is to comprehend the twin life-currents. In the spiritual life, the twin life-currents are accelerated, manifesting a sensitivity. If the disciple has failed to think upon the procreation acts in a sacred manner, he will undergo purifying trials in emotions and thoughts.

In Tribal-Genesis, one reproduces his kind for the strength of the tribe. In Family-Genesis, one reproduces his kind for society. In Lesser Self-Genesis, one reproduces more inventive and individualistic offspring. In Higher Self-Genesis, because one understands the sacredness of the procreative act, he produces offspring reflecting his reverence and love for God.

As one advances in the Genesis Ascent, the chaste use of the procreative power will produce egos of high evolvement. Chaste intent is earned through purity in heart over many lives. As men become more reverent toward the procreative act, greater egos will populate the earth. When love unites the twin life-currents, the pure in heart will beget the pure.

MANTRAM TO BLEND THE TWIN LIFE-CURRENTS

Instruction: Speak once during the day when the Moon is in one's Natal sign.

The twin flames of my being shall blend as one for God.

SELF-GENESIS PARENTS

There are three kinds of loyalty in marriage: loyalty to the marriage and what it stands for; loyalty to yourself and your soul; loyalty to your mate.

In Lesser Self-Genesis may be found many persons seeking a path or a "Way". These are very often parents—fathers and mothers—who are seeking to lead their progeny into the *right* way. Parents in Lesser Self-Genesis, trained in psychology and sometimes in the metaphysical or power path, often inadvertently neglect their children—priding themselves on giving their children the full right of way as to their individuality. They give to their yet untrained children the full use of their expressive minds and emotions. Such children are neglected children, using certain untrained liberties without ethic. Thus, parents who have found metaphysical logic to be sufficient unto themselves, often find themselves with disobedient, irreverent and precocious offspring.

The Western ego is inherently metaphysically inclined, reared in the idea that excelling or success is more important than merit in character. The present resisting youth may be said to be ripe karma for the Lesser Self-Genesis parent in this age and time.

It is the work of all Higher Self-Genesis egos and parents

to present to the young the ethic of consideration. Character earned through former life Family-Genesis morality grace must be reinstated in each life with ethic. All educational procedures should begin with the culture of self-knowing based upon consideration of others.

To be a good parent one should rely upon the proven aspect of discipline, teaching his children the first rudiments and precepts of obedience to Universal Law. Each parent should recognize the individual soul aspect of each child given into his charge, avoiding partiality or bias, using the divine plumb line to keep the unshaped and untried wills of his children within the Will of God.

Mystically-inclined parents produce disorganized, unrealistic households. Untrained philosophically-inclined parents sometimes attribute to their children nonexistent genius attributes. Spiritually-realistic parents see their children as individual egos and know them to be under the Just Law even as adults are under the Just Law. Such parents fulfill their grace by supporting and fulfilling parenthood within the Image and Will of the Divine Father in Heaven.

The Archangel Gabriel is the Initiator in the harmonics of marriage.

8

MIND AND SOUL

> *Our soul is escaped as a bird out of the snare of the fowlers: the snare is broken, and we are escaped.*
>
> —*Psalm 124:7*

The Soul

> *The Dharma is what teaches you; the Karma is what makes you do it. Only when Dharma-Karma slows down can the soul speed up. If there is any religiosity in the nature, one turns toward the Spiritual Path at this time.*

Man is a Unit of Spirit working with a soul.

The soul is supernatural energy overdirected by the Presence and Image of God.

The soul is superconscious energy and movement.

The soul is a flower containing the seed or the Higher Self, leaning always toward the Light.

As the earth spins, so does the soul vibrate and move, throwing off the old and taking on the new as creation.

Human life begins at conception. Soul life begins with the first movement in the womb.

Souls have their human textures through human situations; their refinements and their sensitivities; their gross challenges; and their thrust for life.

A great Great Soul is an Illumined Soul.

One man's victory affects all souls in the world.

No one has the right to own another person's soul—that is the law.

Sins are crimes against the soul.

The consummation of sin is death to the soul.

"And he answered them, saying, Who is my mother, or my brethren? And he looked round about on them which sat about him, and said, Behold my mother and my brethren! For whosoever shall do the will of God, the same is my brother, and my sister, and mother." (St. Mark 3:33–35) Every disciple who crosses into Self-Genesis must undergo a battle between the hypnotic expectancies of his relatives and friends and the holy expectations of his soul. The battle is won when one can observe this play of expectant-forces ordinarily ruling the physical life in the world and the spiritual life of the Kingdom.

Soul is not a wanderer. The soul desires to utilize the Earth-plane. The astral world is an energy-curtain veiling away Heaven to the outer consciousness of man.

Before the Astral World was formed, man lived in an etheric world. It is the work of all men to come alive and awake to the Inner Kingdom called *Heaven*.

Heaven is another dimension of higher frequencies. Every eternity system has a Heaven. Men of the Earth are unaware of the Heaven-frequencies. The dimension of sleep and the soul-dimension within is rarely known by one who is engrossed in concepts, ideas and theories. However, in sleep, man is exposed to the heavenly-dimensions. Were it not so, there would be no hope for him.

The *Soul-Covenant* is learning through the use of the will.

The first Center from which man conducts his human world is the human soul. Unless responsibility is in possession of the human soul, man is uncentered. The essential Esse in the human life is responsibility.

Man is being put out of space because he is not giving his soul space. If soul has to break up total societies in order to give space for itself, in certain periods it will do so. From time to time this occurs in societies which have become godless; in religions which are suffocated by ecclesiastical politics.

Soul, to maintain its passage way, begins its work by demoting the high in place, such as the priestly favored, the monarchy, the dictatorship.

Soul is a calming-machine. Self-passion is a devouring machine.

It is only when you understand reincarnation that you can understand the soul-processes of pre-birth.

If every egotist would accept the fact that he is in the state of learning and earning, he would immediately make a leap forward in consciousness, and thus become aware of his own forming and shaping as a developing consciousness-being. The habit pattern of the egotist develops the Under-soul pattern, reinforcing the ego rather than the Upa-Soul action of the soul.

In this eternity system on the physical plane, it is necessary to experience adversity situations so that one may develop dimensional judgment or spiritual muscle. By commanding the forces on the plane of the physical world, one obtains morality. When the physical-plane force mastery unites with the etheric, astral and mental energy-processes, one becomes a Master in the physical world and in all planes.

The soul of man is fulfilled when he loves God first. All things then are made perfect in love.

The way of life is not attained through another. No one having soul-equilibrium seeks to live *in* or *through* another. He is aware that to seek to make one person all, creates sin —and sorrow comes.

He knows that all seek to live in one another and for one another, yet one must think of the other as having need of union, first with God. Every man must consider the soul of the other. The right of the soul to experience belongs to the soul.

The fine line between caring and loving; the ethic of non-intrusion; the fear of loss of the one loved—these are set before the disciple-initiate in testings, trials and temptations.

Men will come to know: to love a child, wife, husband, friend or animal possessively is sin; to know that one cannot lose what the soul is, is joy; to know that generosity in relationships frees the increases of good, giving health, strength and creation; to seek to take from any one what the soul is, is to suffer an unholy rejection.

The soul has provided seven degrees of memory: (1) the memory of devotion to God; (2) the memory that one should remain pure, chaste; (3) the memory that the forces of Nature are interpreters of God; (4) the memory that one must act honorably; (5) the memory that the thought has imaging capacity; (6) the memory that one must bind up the wounds of his fellow man; and (7) the memory that he will ultimately unite with the Eternal One and create with Him. When any one of these seven degrees of memory within the soul is obscured, the results are confusion, chaos, suffering, pain.

My vessel of Light—the soul—overflows into every portion of my being, giving me the joy of a fulfilled mind, heart and body.

I am happy with a most divine happiness when

I keep my trysts with the rhythmic tides of the soul.

MIND AND SOUL

Thou shalt love the Lord thy God with all thy heart, and with all thy soul, and with all thy mind. This is the first and great commandment.
—St. Matthew 22:37,38

The mind is the most precious thing you have—next to the soul.

The mind is a vehicle for *life* experience. The soul is a vehicle for *spiritual* experience.

Always it has been the spirit and the soul that have resurrected one through the mind. It is thrilling joy to have ignition of the mind and the soul. There is no greater joy in the world.

The soul measures by the Eternals. The mind measures by the memory.

The blood is a vessel for memory. The blood is a vehicle for memory.

The soul has an Eternal mission. The higher mind as an instrument for the soul also has a mission. It is the work of the higher mind to externalize the soul's power and thus build dimensions yet unseen. No one can estimate what the higher mind can conceive. The higher mind has one true gauge—the soul. To trust in the soul's eternal mission is to produce ideas of the Infinite. To believe in the mind as a thinking process for one's own survival and convenience, is to limit the soul's mission. The Divine Intelligible stimulates the soul, and the higher mind responds—creating and expanding.

Supposedly, all men under God are free to make choice. This is often spoken of as free will. Much grief and sorrow, disillusionment and destruction would be avoided if men could truly understand the soul's action under the Will of God. The soul waits patiently for men to comprehend their own limitations as personal self-will egos.

The greatest discovery man can make is the discovery of God's Will for him. Free willing for self invites discipline and correcting. A willing for the Will of God gives freedom of expression under God.

Man is not an automaton or puppet pulled by the strings of an afflicting hand. He is a developing mind, potentially a son of God. The Will of God is centered in his higher mind. Man becomes a creator with God through his higher mind.

To be blessed with a will-surrender to the Will of God is freedom of will. The soul is animated and illumined when the lesser will is made quiescent through an act of godly will within the higher mind.

Soul-techniques are spiritual transformers producing ingenuity of the mind, mastery over the emotions, and victory over the gravity trials of the body. Soul-techniques have been passed down to man by the Greater Avatars, Bodhisattvas, and lastly, most perfectly, by the Christ.

One begins the mastery of soul-techniques by looking within for miracles, by believing on the transforming powers of Transubstantiated Presences.

Self-denial for the sake of the inner life is rewarded by a new mind. The higher mind functions when set free by the divine contemplative reflections on things spiritual.

To heal through soul-techniques is to be healed. To think through divine reflection is to create in timing to man's receptive need. There is a way, a right way, to free one's inner life. One should pray that he may devote himself to the soul-techniques of meditation and contemplation, and thus enter into the transcendental, sacred portals of Reality.

The sweet savor of soul-frankincense gives off the holy aroma of the Angelic and Saintly cloud of witnesses who work with me in mastering the soul-techniques.

The Soul and the Angel-Flow During Mental Initiation

My soul knows and will transmit to my need every rightful thing God desires for me when I use the constant rhythms of my True Self.

Man need not be loveless, for he *is* loved; of all creatures under God, no one is so loved as man. Imaged by the Father, each one is a love-reflection center into which God would come to clarify His Image. The Angels devoted to man's rise love him with unwavering, upholding love. The Saints, knowing their own times in the earth in not feeling loved, love him; they seek to turn his heart toward all of the devoting, protecting sources of bounteous love, particularly toward the Lord of Love.

The Cherubim keep alight the Altar flame where the Love of God may be spoken. The Great Host and Seraphim exalt and praise God's Love. To seek to prove intellectually that one *is* loved closes the door on love. Only the heart can know love and feel when one is loved. The heart then should be kept loving, that love may feel at home in the contented cherishments of being at one where love lives, breathes, blesses and fulfills.

Ignorance of the Angel Flow existing in the mind of man limits one to mundane life. Ignorance of being God-motivated seals away the spiritus flame within the higher unconscious. During initiation to free the higher mind, Angel Guardians watch over the God-penetrable points of the mind. These Guardian Ones work with God to protect the initiate from his own unknowing.

To keep the Angel Flow alive, one should keep in mind that Jesus, who is in command of the initiatory processes of the mind and of the Angels working to give spiritual and divine liberation to the mind, is present, near, and ever responsive to the call from the one in the phases of mind initiation. As Mediator Guardians, the Angels working with Jesus open to the initiate a certain unique logic.

During initiation to the higher mind, the initiate is given a permanent key command-thought to retain in his mind concerning God and the reality of God. This key command-idea or thought enables the initiate to overpower the forces of Satan. From this key command-thought or idea concerning the reality of God comes the power to know God and to interpret Him to all who would hear.

MANTRAM TO STRENGTHEN UNISON WITH JESUS AND THE ANGEL-FLOW

To speak the following mantram constantly will strengthen one's unison with Jesus and the Angel-Flow. Speak the mantram with an upward flame-like lifting of the consciousness. Speak often when the mind is confused, tense, distracted. Speak with ascending, affirming and belief.

Thou Jesus One, let Thy sayings be freed in me.

THE LOWER AND HIGHER MIND

He who rules his mind rules the universe.

The mind travels faster than light.

The gift of the selfless self can be produced only by a pure heart, a pure mind, in Christ.

The mind works best when it is a stilled vehicle.

A flexible heart and a flexible mind make a perfect vehicle for the Presence of God.

Thoughts are the off-shoots of thinking. Thinking is the product of the mind. The mind as imager and knower thinks with the Divine.

This is a versatile age, and a versatile mind is asked of you.

When chastity is offended, the result is mental death.

When you clean up your mind, you clean up your person.

You can have a tossing mind or a tossing heart.

A senile mind is a misdirected mind built by a tossing mind unreceptive to instruction, especially the truth about oneself.

The mind is like a wild horse that wants to charge everything and bring the senses in and have the senses to their full saturation; this is the lower vibration of mind. However, when consciousness is filled in the mind with knowledge of Light and God, then the mind becomes the steed or the stallion driving the chariot of your soul.

The material mind exists and survives on skeptical wave-energy patterns, on arrogant wave-energy patterns. This produces the hypocrisy of finite thinking.

When you give up mentally, you are dead.

Every disease degenerating the cells of the brain, causing schizophrenia and paranoia, stems from sexual deviation and is passed on to the second and third generations of progeny. The sexual sins of the fathers are enscrolled in the brain.

Sexual deviation prevents the birth to the Omniscient Cell in the brain. Drugs taken for sexual stimuli and inciting the grotesque imaging moods in man prove and reveal that genetic stability has been fractured through sexual deviation.

All psychic powers are experienced by man on the level of his karma. The psychic nature works through the lower or instinctual mind.

The spiritual nature works through the higher mind. When one clears away the debris of the lower mind, the dynamic processes of the higher mind can work.

The lower mind lives totally through relativity; the higher mind lives totally through the Transcendental, which means the Ascending Grace.

Truth is the food of the mind, and love is the food of the heart.

CONSCIOUSNESS

The greatest living wonder is Man.

One has a total university within himself when he understands the alternates in consciousness.

When the consciousness is full of its self as consciousness, full of its self as awareness, it is ready to return to the Coinciding; it is ready to use the holy synchronization within the Constant.

Every plane of consciousness has its own laws and governing Spirits.

Your consciousness attracts to you more of what you are.

If you are thinking failure in your consciousness, you will automatically attract it.

When reverence is absent, and the intellect is too much, the spiritual cannot work.

Force: a congregation of energies either in conflict or in a processed state of another degree of energy.

Energy: seeking to make more expanded processes which consciousness can draw upon.

The consciousness mind works in a circle. The etheric or

inner mind works in a sphere. And the spiritual mind works in a spiral. The outer produces appearance. The inner produces likeness.

Mind is the visualizing power of consciousness.

THE EGO

Beyond the masks of countless egos experienced in former lives is the True Self, the Real, the One in the One.

The *ego* is the vehicle for the intellect.

The ego is the number one idol on the pedestal. The lower mind and the lesser will will do anything and make any sacrifice to save the face of their ego—their idol.

Everyone who is ego-bound is lacking in reverence.

There are some persons who are not on the Path; they are *exposed* to the Path. They have a two and one-half year exposure to the Path. Then the ego takes over and diverts them into the Undersoul darkness rather than the light of the soul. They are limited to a two and one-half year endurance-cycle.

The bigger the ego, the smaller the capacity.

The shaded side of the ego is like the dark side of the moon.

There is a place in the subconscious which does not cheat or let one cheat. There *are* places in the subconscious which do cheat, and thrive upon fantasy and self-intoxication. He who lives in this cheating-part of the subconscious is a dreamer upheld by self-untruths and ego-inflation. One must go to war with this cheating level of the subconscious. He must ruthlessly cleanse it out with the two-edged sword of Honesty.

The greater the ego, the greater the agony in its death, and the greater is spiritual power in its resurrection.

Holy longing must come before the ego becomes the server and servant for the True Self. Holy longing crosses the bridge of fire into Light.

THE TANMATRAS AND THE THINKING PROCESS

Love is the unifier, the blender and the salvation.

Each sense is connected with an activity of the mind, the emotions and the body. When the senses become thought as activity, the thoughts clothe themselves in one of the elements. This is tanmatra in action. The tanmatras are five; they are air, fire, earth, water and ether.

When a thought is clothed in tanmatric fashion as fire, one expresses thoughts psychically, and is obsessed with a burning passion which sets fire to his life and to the lives of others.

Any of the tanmatras as thought can cause obsession: as a thought clothed in the tanmatra of air, one is scattered, uncoordinated; clothed in earth, one is unteachable, tamasic; clothed in water, one is unstable; or if one is in a pure state of water-clothed thought, he is a healer and a purifier through thought. If one is in a tanmatra state of ether in thinking, he is thinking through Angelic Coincidings and is telepathic to all levels of the tanmatric thought action.

Racial thinking, when aggressive, uses the fire or psychic tanmatra. Obsession then occurs as to racial reproduction of thoughts and ideas supported by racial prejudice and separateness.

The tanmatra miracles occur when one saturates his thoughts with the ethers of the Coinciding. One masters the tanmatras when he has achieved spiritual power and adeptship in the use of the tanmatra-energies as produced in thought.

This explains how thoughts clothe themselves in elements. When thoughts are negative, this causes obsession in thought or mind.

Every sense correlates to a tanmatra. And every sense has a function of thought to the making of thought processes.

One can literally define the evolvement and evolution in thought process by what sense predominates in his thinking: seeing, tasting, touching, hearing, smelling, or lastly, through the etheric side of his nature.

One must clothe his thoughts with the tanmatra of ether before he becomes a ministering adult in the mind.

Desire and the Path

> *We are all frail vessels set upon a strong journey. We need to remember that Resurrection Powers were given to all men.*

Soul-dependency is healthy between persons. Soul-dependency is a true yoking and co-atom experience. Desire-dependency is cruel, separative, and disconcerting to one who must climb over his desires for supports built upon sentimental expectations in the desire nature.

Sentiments are chimeras or longings unfulfilled in the desire nature. Hardship-karma relates itself through false sentiments within the desire nature.

When we desire that our young become worldly representatives of our ideals, we short-circuit the child's soul-dependency nature. One cannot live peacefully with himself or his conscience as long as he seeks to *will* for others. No matter how much we love someone, we are not in control of their destiny.

One should mark and trace his desires; search his motive-thread within his desire. He should also stand back through Marking and Tracing, that he may permit the inter-soul flow to be experienced by those whom he loves.

To accept the soul-flow in others is nobility, integrity. To seek to stifle, twist, or direct the will of any person as to his desire to worship or his desire to create, is to thwart the soul-atmosphere of others.

In the Path of Initiation, there are crossing paths to one's own Path. One must learn to recognize these intersections containing a cross-walk energy which can divert him from his true Path. One is called a *cross-walk one* when he reaches these intersections where he is challenged by someone on another path leading in another direction. This crossing point on the Path would lead him to everything his soul does not desire for him.

A Teacher who has earned co-atom grace becomes the directing energy to help the chela put his character, intent and motives into the integrity channels. The Teacher firmly points in the one direction toward the soul. Thereafter, the seeker or one desiring shall become co-atom to the Teacher, that he may see eye to eye the Path as the Way for him, that he may become co-atom to the Lord.

On the level of desire, it is men with ideals who support the ideas of those who are born with co-atom Niscience or Knowing. With the beginning of insight, it is the first duty of all to channel their idealistic natures through service to God. To recognize a God-sent person is holy insight.

Everything in this Earth comes as an idea which has its origin within the Greater Archetypal Energies within the Mind of God.

Those in the world having disorganized desires cannot comprehend Creation in its higher aspects, nor can they understand the soul-constancy within created things existing in the physical world. Failing to have an insight into the power of praising and blessing all forms created and all forms yet to be created, they are disconnected from the soul-flows of creation. Their existence is dependent solely upon the laws of evolution within the involuntary systems of Nature. They are thus animated to be conscious on the plane of lower Prakriti.* Being victims of their own limited energies, they suffer the hardship-karmas of the karmic encasements where men do the same things, thinking them not to be the same.

The desire nature, when mastered, understands fully the system of Maya,* and interprets it as a holy privilege to work with God.

> *Co-atom wisdoms create holiness in environments, in mind-stature, in integrity.*

THE BEATITUDES AND THE OMNISCIENT CELL

Blessed are the poor in spirit: for theirs is the kingdom of heaven.
Blessed are they that mourn: for they shall be comforted.
Blessed are the meek: for they shall inherit the earth.
Blessed are they which do hunger and thirst after righteousness: for they shall be filled.
Blessed are the merciful: for they shall obtain mercy.
Blessed are the pure in heart: for they shall see God.
Blessed are the peacemakers: for they shall be called the children of God.
Blessed are they which are persecuted for righteousness' sake: for theirs is the kingdom of heaven.
Blessed are ye, when men shall revile you, and persecute you, and shall say all manner of evil against you falsely, for my sake. Rejoice, and be exceeding glad: for great is your reward in heaven: for so persecuted they the prophets which were before you.

—St. Matthew 5:3–12

> *I have not been able to see any difference between the Sermon on the Mount and the Bhaga-*

*In Sanskrit, Prakriti means *Nature*; Maya refers to the world of *Illusion*.

vad Gita. What the Sermon describes in a graphic manner, the Bhagavad Gita reduces to a scientific formula. It may not be a scientific book in the accepted sense of the term, but it has argued out the law of love—the law of abandon as I would call it—in a scientific manner.

—Mohandas K. Gandhi

The Ethic of Jesus is concealed in the Beatitudes. Jesus laid down the Ethic for the life of the soul in the world in His Sermon on the Mount and in His reminders of the necessity to live within the Commandments.

The Ten Commandments and the Beatitudes are the vital remembrances. They keep alive the true flow of remembrance between man and God.

The Beatitudes are the Great Compassions.

In the birth of the New Era, when men will look toward a world and earth made pure by return to Principle, there will be egos lifted to leadership through the use of Ethic based upon the Sermon on the Mount. Before men can enter this time, there must be heart and conscience research.

The nine openings in the body relate to the nine Beatitudes. The tenth opening is to be in the heart ventricle. As the organ for the Supreme Self, it will move toward the center of the body of man. This will be the tenth opening. The centering will be considered to be the opening. This will coincide with the perfection of the etheric brain through which the *Omniscient Cell* can function in the use of the higher mind.

In the New Testament, the Beatitudes contain the Ethic of Jesus for the use of the Omniscient Cell. As the Laws of God were sealed into the inward parts or chakras of man, the Ethic of Jesus functions through the ten openings of the body. In this era and time, all Cosmos Disciples use, to some degree, the Omniscient Cell, which sustains their physical body until their work is fulfilled. The Omniscient

Cell also keeps them in alignment with the Greater Archetypes through which the Word of God shines as a golden splendor awaiting to fall into the minds of the prepared.

Today's creation is tomorrow's preparation.
My soul shall harmonize the Beatitudes
and the promise of the new day;
and I will be born again.

CONSCIENCE AND THE GUIDING PRINCIPLE

And herein do I exercise myself, to have always a conscience void of offence toward God, and toward men.

—Acts 24:16

The chief agent for the Guiding Principle is your conscience.

Let your conscience be listened to at all times.

If there is anyone among you who is leaning in the direction of adultery, let him listen to his conscience.

If there is anyone among you who is coveting, let him listen to his conscience.

If there is anyone among you who is not telling the whole truth about himself, let him listen to his conscience—and his True Self will come forth and speak to him. And his conscience will be shining as a sun, as a light of Heaven.

Personality-love relates to morality; soul-love relates to the conscience.

People with a confused mind do not have direct access to their conscience.

It is the conscience that makes every sickness in the body. Were it not for the conscience, man would have no sickness.

There are millions of men who are letting their wives be their religious conscience.

A prophet is a first pro in the jungle of the human conscience. As a pro, he is taller than his adversary, untruth;

seeing over and beyond; dismantling the lies which have entrapped the inert and the lazy.

The heart is the seat of the conscience.

ARCHETYPAL REALITIES

Jesus left the Commandment-of-Love Ethics for us to prove. This is the joy of realization and expansion through the love and the mind.

The *Archetype* of Niscience, as a new spiritual and mental vibrancy, provides a stimulating, inspiring discovering into the combined souls and minds of all. Man is presently activating soul-powers of the mind never before encountered in the world.

From this universal downpouring of Knowing, men will probe, research, and learn of life, life's creation, life's intent. With a universal perception, all souls will be plunged into command of little-known forces, energies, and new dimensions. Newly-expanded values of the mind will draw upon levels of consciousness akin and close to the thoughts of God's very Elect.

Advanced souls over the ages have anticipated this great eternal day of mind ripeness, having moved the flood tide of mental and spiritual grace from life to life. Such men look toward God to maintain their new spiritual perception, that they may uphold, teach, and enable those less evolved in the world to keep their minds fixed upon the realities existing in and of Universal Law.

God's Laws of the Universe, set into action from the beginning of this eternity system, watch over and keep balanced all growing, developing and evolving life. God's Universal Law works to maintain and sustain all men. No matter on what level of evolvement anyone may be, all will benefit and progress in this Archetypal Age of Niscience or Knowing.

As a responsive infant responds to the hopeful parental

love in each new life, so shall those who are hopeful, trusting and loving respond to the new order of the New Age now opening toward ripeness of the mind.

A System of Archetypal Metaphysics is now being established—a System containing a blueprint for a fearless mentality, a divine creation and envisioning mind.

Hierarchal powers of the mind are universal rather than personal, limiting powers of the mind. Men now stand nigh to the threshold of Hierarchal powers as promised by the Hierarch of this eternity system, Jesus.

That One whom we call Jesus came to this world to prepare all souls for this eternal day—the day of man's coming of age to the Mind in Christ. To look on men as they appear, to think on man as impotent in the sense of things spiritual, is to seal away the realistic truths as established by Jesus.

All men, in spite of their fears, doubts and unknowing, are now facing the birth of a new mind. The flowing tide of *Knowing* is at hand. He who holds the key through Jesus has opened the door to a new and wonderful world of mind creation.

The mystery of my beginnings resides within the record of my soul. The mighty Eternal awaits to speak and reveal to me the I AM THAT I AM.

9

GENE-INHERITANCE

A good man leaveth an inheritance to his children's children: and the wealth of the sinner is laid up for the just.

—*Proverbs 13:22*

Genes and Chromosomes

The Sun is the energy used by Hierarchy to stabilize the chromosomes and balance them. Hierarchy works with the chromosomes. The Planets, the Moon and the Angels work with the genes.

The higher and transcendental psychical impulses are sealed into man through the chromosomes. The ancestral canvas upon which one paints the story of his soul's sojourn is centered in the genes.

The little understood pre-birth psychical selective soul-power enacted during the time of impregnation and ovulation between the sperm and ovum of the father and mother—choosing and directing the particular gene to give embryonic life to the soul—is not an accident. The choice of genes is a decision made by the ego whose karmic pattern and selection determine the color of the eyes, the skin texture, the physiological and functional organism of the body, frame, stature, and chronic tendencies in health. In the moment of copulative gene infusion, all of the things related to ancestral karma and ego karma or soul-grace are decided by the soul entering the world.

The number of active chromosomes working with the genes during conception is determined by cosmic equation. The genes determine the ego's fixed field of fate through which he must overcome certain former-life and inherited ego-tendencies of moral and physical weaknesses while living in the world. Balanced chromosomes come under the influence of the Celestial Elders or Hierarchy determining the potential of man's fulfillment in his spiritual destiny.

Until the coming of Christ through Jesus, man lived totally within the involuntary system. With the Passion of Jesus, and the opening of the Indestructible Atom, man began to create himself. Consequently, all sins of mankind caused by offense against the chromosome system and the gene system must be transversed or mastered by the infinite processes of unfoldment through the self.

The Yantra Initiation experienced through the Mouth-of-God Center at the base of the brain opens one to the myths of all cultures, which hold the clue to present-life inheritance from past lives. The genes contain the myth-memories of ancestral flow; the chromosomes, the myth memory of Hierarchy impressionability through the myth-flow.

The Angels are the guardians of the genes, and the Hierarchs are the guardians of the chromosomes. The Angels, as guardians of the genes, work through the ancestral flow of myths related to families and generation cycles. The Hierarchs work with the chromosomes in the forming and shaping of man.

Everywhere there is the evidence that certain men are responding to the true and basic myths of the past and also that mankind as a whole is being prepared to develop a new or virginal myth symbology for the future.

GENE-MEMORY

Gene-memory contains both virtue and darkness.

Gene-memory is man's inheritance through which he grows to mind and soul stature within the Increase of God.

He who seeks to transenergize gene-energy memory can do so only through spirituality. It is the Law of God that the sins of the fathers are erased by the progeny of the fathers. The sins of the fathers are overcome though spiritual transenergization processes. In every Family-Atom, souls incarnated together are united to increase their virtue and to master the negatives or the sins of their fathers. In every third generation in a Family-Atom, there comes to birth one soul who has the power to transenergize the sins of the fathers.

The seer as a prophet crosses the bridge of the lower subconscious, which relates to gene-memory and the reincarnation soul-processes of the past. Having mastered these, he walks beyond the abyss into the Bridge of Light where he makes union with his hierarchy nature and becomes an imager-creator for God.

The knowledge of the Law of Reincarnation and the understanding of grace and karma will enable men to use to the fullest their inherited propensities. Those who truly understand reincarnation understand also that an ego consists of many loaned and borrowed attitudes in the mind and in the emotions. And in true understanding of reincarnation, the soul is clearly known and utilized from birth to birth.

To be wrapped in the blanket of genetic limitation devoid of soul-expansion is to be little more than an insect, an animal, a tree or stone.

All initiates have the power to break the inherent genetic karmic chains; however, if untrained or faulty sentiment is present, one is subjected to intermittent genetic karmic possessive intrusion from the ancestral dead and from the living.

The Undersoul is carrying the greatest ancestral-gene karmic burden it has ever carried in history. This is the cause of man's upheaval and the distortion of his conscience. The pressure of the gene ancestral record comes that men may be cleansed in preparation for the Self-Genesis Age.

Yin and yang, during the period of the nine months in the womb, imprint upon the one-to-be the identification marks as to appearance, coloring, perfection in functioning or karmic malfunctioning in appearance. Appearance is a yin and yang work. If there is ancestral karma, appearance will follow the negative pattern in the genes inherited from the parental line.

When one is born, he automatically inherits through his genes the sins of his fathers, even "unto the third and fourth generation." (Exodus 20:5) To the novitiate or disciple, this means that when he overcomes the tendency to sin, he frees four generations of ancestry who have lived before him. This is one of the ways he honors his father and mother and his ancestral line.

When a person is highly evolved, he is less likely to duplicate ancestral weaknesses in his physical body. When a person has lived in sentient-atom experience in former lives, he is more likely to inherit the physical weaknesses of his ancestral stream. Mediative healing may enable a person to rise above ancestral inclinations and tendencies. The Mediative Healer should seek to reveal the true ancestry to such persons—that is, all men are created in the likeness of "our Father which art in Heaven." Thus, our Father is the true Parent or Ancestor.

Fear, acting within the low-octave of karma, leans one in the direction of gene-tendency in disease.

All restitution sicknesses relate to chronic, crippling diseases in which the bone structure is distorted. This relates to former-life gene ancestral-karmas inheritance. Restitution Grace enables the sick with bone diseases to unburden their ancestral inheritances. Through service to the human race, Restitution Grace is made.

In this latter part of the Kali-Yuga darkened time, those who reject their children—whether by abortion or by birth itself—are self-destruct in their own life-force and destruct in the life-chain of gene regeneration. Every woman who willingly or ignorantly aborts a child has rejected a soul-presence in her life. At some time, the woman who aborts her babe is compelled through karmic law to give birth to the same soul she would have borne.

> *When you dream with regularity of the gene-line, you are working mediatively with everyone's gene-lines. Gene-line dreams occur mostly on Fridays or Saturdays. It is in such dreams that one tests himself also to see "who is my mother, sister, brother?"*

GENETIC SELECTION AND THE SECOND CHAKRA

In the two thousand years since Jesus, the second chakra of the Kundalini fire of mankind has functioned in conjunction with the accelerated fiery *void* within the core of the earth. Before Jesus, the genetic flow of sexual life in the second chakra was under the command of the Race Lord Jehovah. No longer under the total influence of the Race Lord Jehovah, man has begun to individualize the ego beyond race. This is opening him to a higher aspect of the unconscious, making him more directly aware of the universal cause rather than physical reactionary cause. With the coming of Jesus, the Self-Genesis man began to function within the etheric matrix of the suggestible Archetypal blueprints. The seeds of Self-Genesis have taken root in the higher unconscious. Man, now more and more as an individual, will seek to unite with a third aspect of soul-power available to him.

In ripeness to the soul's timing, one moves toward spiritual birth through the mastery of the second or Pelvic Chakra. Through the use of the spiritual practices, powerful mantras and mantrams, and service to God, one finds the joy of freedom from gene-memory, gaining the delights of Divine Selfhood. This is a delicate and sensitive time which otherwise can be glamourous and deceptive to the seeker if there be any remaining form of sensuality.

The astral portion or disc of the Pelvic Chakra is a repository receptacle for the lower subconscious. The gross or crude planetary forces of Saturn and Mars keep alive the memory-chain of the inherited ancestral negative-sexual actions. These stimulate, from time to time, the ancestral memory of guilt, weaknesses, and family traits which are passed on through gene-karma.

In marriage on the second or Svadhishthana-Chakra level, Yang in man looks for a mate through the genetic and Genesis selection, seeking by family conscious instinct the right hormonal balance to produce offspring of strength and normalcy. Woman, as Yin, seeks subconsciously a mate to mother and to give her the seed for her womb according to type selectivity in gene balance. By instinct, the woman on the second-chakra level of marriage initiation represents the Divine Mother; she holds the key to love and sacrifice, assuring the rise of the human race.

Yang male makes his selection in marriage through soul-to-soul recognition of his mate. This Chakra-plateau of Yin-and-Yang rapport may be likened to all lovers of history who have fulfilled the ideal of gene blending and soul integration.

Jesus—the Cosmos Avatar bringing the Christ Mind to

the advanced men beyond the fixity of the Jehovah-karmas within the second chakra—came to produce in men the way to obtain sustenance for survival, that they might live beyond the instinctual, aggressive-will memories of their ancestral recollections. He came to unite the souls of men with eternal recollection, to remind them that they are eternal souls covenanted to embody in the world to sensitize the gross vibrations of the earth, and to use these vibrational forces to produce the ultimate paradise in their own natures and in the world.

Heaven will be nigh unto my call if there be light in my mind.

ANCESTRY

Worse than the meta-circus is the gene-circus.

In the breeding or reproduction of man, the complexities in grace and karma being unidentified in the present consciousness of man—and reincarnation being little known in the Western world—men will continue to be deceived in the understanding of inheritance and of gene succession as given to the human soul.

In the Maya system of this earth, one could liken himself to a fish in the ocean of Maya, engaged in an inevitable pursuit deemed necessary by blueprinted ego habit-patterns and inherited gene-patterns passed on by ancestors to their heirs and also printed upon the lower etheric encasements as subtle habit-patterns followed throughout many lives.

The subconscious is a recorder and a regulator of the illusionary aspects of Maya. The superconscious is Reality. The outer or objective consciousness is a sense-limited vehicle expressing more often the mechanized processes of gene-inheritance.

No ancestral pattern in the cell of man can compare,

challenge or make void the record of a past life. The record of former lives holds the preponderant or superior place over any ancestral inheritance. What man has been in other lives determines what he will do with his gene and ancestral inheritance, how he will relate himself to them. If he has more soul-memory of eternal inheritance, his auric light and auric vehicle will begin its blessing on the first breath of his living. However, if he brings to this world a record of violence, hate, hostility, he will unite with all of the negatives of his gene-inheritance, and he will scourge the earth rather than bless it.

When one is born to the Maya scene in the physical world, his Undersoul inheritance covers him as a cloak, sealing him into gene-memory of good and evil. His own ego-memories of good and evil set the scene for the drama of the years facing him in the world.

Every man enters this world burdened by Undersoul contamination which he has written into his account of life in earth. The record in the Undersoul inevitably calls for payment of karmic debts incurred by one's ancestors and ego in former times. One must battle the memory flowing from the attitudes and prejudices of ancestors.

Jesus makes each man a steward-processor in the use of grace, thereby making him accountable to the Law through the Commandments, freeing him from the bondage of heritage-karma through the genes and through sin. When one takes the Continuum Sacrament, God through Jesus resurrects the misused talent and neglected grace which is the natural heritage of all created by God.

Every act of virtue one would seek to attain is sifted through the screening process of past-life grace and past-life karma. If there is more grace than karma, one enters the world unburdened, free. If there is more karma than grace, one assumes his ego-karma and the karma of his ancestors at birth; and, throughout life, he must weigh the works of this life and works of other lives against the scales of grace. If one has energized his grace in a past life and has built a reservoir of grace, he operates and functions in a new life without the compounded chemistries and biased outlooks of his ancestors and ego of the past.

To be grace-born is to be destined to fulfill an aperture in the Maya chain as a grace-blessing to all.

In every ancestral inheritance, one is fortified to some degree by some strong virtue existing in the ancestral gene-memory. If there is ego-strength, one brings into the world an ancestral strength; this is birthright-grace, enabling the one entering the world to fulfill the higher ideals which have been manifested in previous lives.

Many egos born with weak life-force resolution, having good ancestral or gene-memory, fail to activate the strength of their ancestors. However, such egos, having progeny, can pass on to their grandchildren character-strength which they have failed to use. Grace invariably marries strength of soul, mind and body to one another; in this, a grace-destined ego is born in unusual environments with unusual opportunities.

When the Holy Ghost works with the tones of propagation, He has the power to change fixed patterns of generation. The tones of the Holy Ghost, when necessitated by vital spiritual urgencies, can change the tone-combining between sperm and ovum. A soul to be born into a pure life is assisted by the Holy Ghost to accelerate the supernal etheric vitalities within the genes of his parents; therefore, he enters the world with a chaste body, emotions and thoughts. This is a sacred birth, a holy birth.

To tamper with or disarrange the genes through physical means is a sin against the Holy Ghost. The human race is in danger of becoming degenerated when men seek to exploit or to manipulate the forces controlling birth and death. This is the work of God and the Holy Ghost. There are bacteria that man has never seen, and there are supernal energies and vitalities that man as yet cannot comprehend.

These energies and vitalities are used by the Holy Ghost so that life may be more than bread alone.

Science will discover in its research of the genes the necessity to retain the mysterious energy particle assuring survival. To break the chain of survival strength in any person is to outrage the Will-of-God compulsion existing in all sentient life in earth and in all worlds. When one closes the door on reverence for the source of his physical life, the survival impulses are reprocessed into destruction.

To retain the will to survive in sickness, one must return to love for those who gave him a physical form. One should seek not to worship the source of his physical origin, but to look with awe and acceptance upon those who struggled to rise that he might someday enter the world.

One seeking to open the door to the spiritual life should keep his mind open to the strengths and the inherent weaknesses in his human nature. He should assess his ego inheritance as being more than his ancestral inheritance, yet he should also acknowledge that he lives in a human world outpictured by countless relationships all interdependent on strengths and weaknesses. His soul and the higher aspects of his ego will come in time to unite him with his true Parent, our Father which art in Heaven.

> *And for this cause he is the mediator of the new testament, that by means of death, for the redemption of the transgressions that were under the first testament, they which are called might receive the promise of eternal inheritance.*
>
> *—Hebrews 9:15*

GENERGY

The One-Constant is Spirit. The vehicle for the One-Constant is Light. The functioning of the vehicle of the Spirit as light is *Energy*. When Energy works as sentient form through mass, this is the spirit made flesh. Energy as density through flesh manifesting as consciousness is *Genergy*. Through the higher-mind energy expressed in the genes of man, the parabolic and archetypal-symbol functions desire to be utilized as revelation and creation.

Genergy, or gene-consciousness, functions first as biotron energy; second, as unitron; third, as omnitron. All of these are developing phases of genergy-consciousness.

Genergy is the gene-patterns inherited from one's racial and ancestral compulsions.

The disciple must be grateful for what his ancestors have done for him, for his ancestors have prepared for him a genetic-soil through which he must come to the earth and experience the earth.

There must be a survival of the grace-memory in the race so that one may attain in the overcoming the lower side of the ego.

Mass schizophrenia from conjunctions of ancestral karma, conjunctions of traits, produces conflict.

There is demonic, entitized karmic energy in a race and in a family. This is the *Dweller*; it is unintelligent. To be exposed to this demonic form is to undergo the most crucial of initiations. Only in the very last stages of dying to one's own nature within the egotistical shell is he exposed to the demonic dweller. This happens in the life of one who is to give the whole of himself to God. Both men and women go through the dark Kali experience.

THREE POINTS OF LIFE WITHIN THE CELL

> *All cells and organs of the body are in motion. This unceasing motion correlates to the movement of the Cosmic System supporting all organic life, and to the rhythm of the Cosmos System supporting the spiritual life.*

All health begins in the etheric body. Every organ and every cell of the physical body has its superior overlord or an etheric counterpart.

To be indifferent to the laws of Nature, to refuse to adapt and to use Nature's laws, to seek to detach one's self from Nature's vital restoring gifts, will set up armies of alien bacteria in the ether supporting the cellular system of the body. The result will be devitalization of the blood system controlling the cells of the body.

The etheric energy in the lesser etheric body, the heartbeat, the pulse points of the body, the cells, the veins, the nerves, the arteries, the muscles, the glands and the organs of the body are utilized according to the mental and will intensity of the ego. The blood as a fluid agent is a photographic field registering man's impressionable emotions and thoughts.

In dream or night initiatory instruction, the initiate is taught to observe, scrutinize and study the cell life in the blood stream. He is shown that each cell contains three points of life working in a pyramidal or triad action. He studies the cellular balance between the three points of life within the cell. He watches the movement of the cell in the blood; and he is shown through initiation how the cell maintains its exchange points of balance, and how the cell in each organ has a *higher* or *spiritual* function, an *ego* or *individualistic* function and an *ancestral instinctive* function. He notes how the cells of the liver, the heart, the spleen, the brain and all other organs respond to a tone suitable to each organ. He sees how the Sacred Atom in the heart maintains the tone within each cell. He also sees how the Master Atom in the heart over the cells, when out of focus or deranged, imbalances the physical body. Degeneration in the cells is multiplied; ether supporting life in the organ collapses, the cells becoming sluggish, weak, and sometimes malignant.

The organs of the body are so constituted that each organ

works in a positive or negative polarity to another organ. For example, the spleen and liver are antagonists to one another; the pancreas, to the adrenal glands. When the various antagonistic functions are balanced in their polarity function, there is health in the organs and in the body. The cells also work in positive and negative functions in their triad points of life.

The Sacred-Atom tone within the heart commands the negative and positive actions of the organs and of the cells. The Sacred-Atom heart tone also commands the rejecting rhythm of the cells, that the old cell may die and a new cell take its place.

If the prototypal image supporting the ego in the Sacred Atom of the heart moves out of accord with the rhythm of the Sacred-Atom tone, the triad cellular function becomes a purging fire. Inharmony between the prototypal image and the Sacred-Atom tone of the heart is caused by a continuing desire to escape from life, or a continued wish to die. Through initiation the disciple learns that persons having such suicidal wishes lose touch with the coordinated rhythm between the cell, the organ and the prototypal function of the ego.

Vital cells sustaining the norm of health in the body work with the blood in a continual flowing state of movement. Cells moving with the blood stream are weighed and assessed infinitesimally by the Sacred Atom in the heart where the ego prototypal pictures gathered from the three points of life in the cell are recorded on the prototypal image in the heart. Should the mirrored picture of the cell reflect the ancestral cell as being predominant, the Sacred Atom records this emphasis onto the prototypal image. Here it is registered that it may be retained. Should the cell mirror the picture of over-accentuated activity in the individualistic point of life in the cell, the result is recorded upon the etheric, cloudy substance of the egotistical shell. The portion of the egotistical shell surrounding the etheric encasement of the heart contains all of the self-willed actions in this life and in former lives.

When the lowest or ancestral point of life in the cell is predominant, the health of the physical body is subjected to inherited ancestral sicknesses. When the emphasis is upon the individualistic point of life within the cell, the physical body may undergo many undefined or undiagnosed states of sickness. If the picture of the spiritual point of life in the cell is inactive, the picture fails to mirror itself upon the Sacred Atom of the heart; this is due to one's having depended solely upon the instinctual will rather than upon the spiritual will. One is thus shut away from the spiritual flow of the soul.

In the study of the cell through initiation, the disciple will in time perceive the answer to the riddle of all cellular structure.

In the latter part of the Self-Genesis Age, a fourth point of life, as yet latent in all cellular life, will become active in the blood, tissue and organs of highly-evolved men. This fourth point of life in the cell will correlate to an Omniscient Cell in the brain. The functioning of this Omniscient Cell will produce the spiritual man.

The memory in each cell within the blood is divided into three states of functioning: Ancestry, Ego and Soul. Man experiences these three functions of the cell and cell memory in the complex cycles of change within changelessness.

In each cell there are three points of sentient relatedness. The lowest tonal rate in the cell is from the gene-inheritance. The secondary tone in the cell is the individual ego-memory and Undersoul compulsions. The third cellular tone, when reached, is the healer of the two lesser, secondary tones in the cell.

The battleground presently existing in such large measure against malignant diseases is an inheritance of many generations passed down through the lower tonal cell-gate of the genes and the secondary cell-gate of the ego. He who knows the sattvic and practices the sattvic with his hierarchy nature is a healer of first magnitude, a master over the tumults of separation and dysfunction.

There can be no health in the body when aggressions are wrongly channelled or misused. The very cells of the body working with the memory system of the will to live, the will to know, and the will to create—when out of balance—produce irritation, inflammation and infection.

Every cell of the body is aggressive. When absent from the sattvic principle of love, there is sickness to the body. When one has finally reached the state of true self-awareness, his body functions in a dimension of health, unity and harmony.

A malignant disease reveals that one has a portion of the mind which is malignant. One thought, when statically fixed within the thought-making process of the mind, is obsession. Malignant diseases are the outer manifestation of one obsessed thought standing before the soul. The magnification of the soul goes down deep into the cesspool of obsession, reproducing the mind malignancy on a certain level of the cellular system.

One should learn to read his own Recording Angel's record, seeing himself as an impersonal, energized vehicle of moods and aggressions. He must learn the difference between the kinetic pressures within the cellular nature. He must see in himself the supersensible inclinations.

If one's aggressions incline him toward intrigue, secrecy, manipulation, he writes his own epithet of existence, for he exists rather than lives, domiciled within ignorance. Life processes to him seem accidental, fortuitous.

Cancer is a cell disease. A cell disease is a gene disease caused by thousands of deviations inflicted upon the rhythmic and cyclic timing in the cell.

The cell is equipoised by the body-energy Law. Sickness in the cell level is deviation from this Law. That which is deviated in Law must be corrected by Law. Willingness to accept Law brings health to the cell, to the blood, to the mind.

The Continuum Sacrament, on quickening the blood through virtue-faith each morning, enables the person to use another rate of blood vibration superior to ordinary physical survival. The quickened blood contains the Archetypal Dharma Light of the Christ Mind, regenerating negative thoughts and negative actions into spiritual associations and accomplishments. The cells in the blood are filled with memory-indexes from one's past instinctual climbing nature. A new vitalizing memory is built through the positive use of the Continuum Sacrament upon awakening each day.

THE SEXUAL LIFE

> *All persons having the inward and unceasing praise and gratitude to God manifest a powerful life-force in harmony with the Laws of the Universe. Such persons appear before others in the world as having qualities to be emulated and desired.*

The Greater Archetypes beginning to flow into the minds and souls of advanced men and women seek to produce in this age a cleansing of mind and body. Due to the slowing down of the magnetic belt around the earth, certain forms of sterility and mutation will begin to manifest after the year 2000 to change the fantasy sexual-glamour presently afflicting mankind. Before this century ends, tremendous famines, floods, plagues, wars and calamities will besiege the earth until men return to the essential chastity-laws sealed into their inward parts or chakras.

When pure love is absent from the act of sex, a form of

anti-creation is manifested which cannot ennoble man or mankind. The primordial Edenic chastity-laws are profaned and defiled when amorality and lust predominate in the sexual life.

A pure sexual life must be supported by reverence, sacrifice and right giving. When parenthood begins as the natural result of love in the sexual life, giving begins. When one thwarts the giving side of parenthood through selfish desire—using the sexual force sensuously and selfishly—he closes the door on the pure and grace side of the genetic flow seeking to be manifested in the world.

All forms of cell malfunction are caused by astral psychical viruses under control of the Destroying Archetypes which are presently attacking the immunizing life-force energies of mankind. In primordial times, these energies were sealed into the thymus gland of man as survival life-forces providing healing for the physical body.

Man has upset the cell, virus and bacterial system through his lack of reverence for life. The outer pollution in the earth's atmosphere is but the projection of man's own second and third-chakra pressures of mixed-up greed and deluded fantasies regarding self and sex.

All venereal diseases are caused by lust and irreverence in the sexual act. Sexual diseases are the retaliative curse placed upon non-chastity actions. Excessive use of sex; demands of unlawful domains and territories of space for selfish expansion; abuse and failure to keep love-promises—these exist as bundles of negative astralocity within the lower planes of the astral light influencing the third chakra.

The Sodom-and-Gomorrah mind thinks through inversion or inverted thinking. Inverted mind produces perversion. In the practices of sexual perversion, all thoughts flow downward through the rectal area or anus rather than upward through the heart as conscience and through the higher mind within the light of the soul.

Degeneration or amorality on a wide scale occurs when a

civilization is expressing the Destroying Archetypes. In the state of civilization decay, retarded egos who have abused sexual purity in former lives, having upset the yin-and-yang sexual polarities within the pelvic chakra, are born to the world.

The perverted person is cut away from the general or universal conscience supporting life in earth; he or she is like a meteor heated with an intense and searing fire, breaking away from the earth body and soul body of conscience.

The homosexual follows the subnormal or *Pan* impulses. The homosexual nature is influenced by the lower-astral Satyr planes where dwell the Pan creatures or elementals.

The Pan dominion has never incarnated in human form. They are psychically communicable to the homosexually-inclined person through his inverted sentient and sexual nature. The Pan elementals accelerate their pressures upon the homosexual during the darkened phases of the Moon; also, in the horoscope of a homosexual may be seen the karmic affliction between Saturn and Venus.

Peace must come. All advanced egos work that a sattvic peace stemming from pure desire, pure motive and pure attitudes may come to earth.

The pelvic or sexual chakra is the coordinate with the fifth chakra expressing speech. The sounding of a prayer, a mantra or a high-praising mantram to God cools, purifies and cleanses the lower-subconscious heated sexual drives, giving chastity-control within the pelvic chakra, the heart chakra, and the chakra seated in the hollow of the throat. Praising God continually will produce a truth-joy in the Self, which will manifest in the world and in Heaven as a recognition and acclaim of one's true and perfect worth.

The astral portion or disc of the second or pelvic chakra is a repository receptacle for the lower subconscious. The gross or crude planetary forces of Saturn and Mars keep alive the memory-chain of the inherited ancestral negative-sexual actions. From time to time, these stimulate the ancestral memory of guilt, weaknesses, and family traits which are passed on through gene-karma.

When the Moon is full, the subconscious is more likely to reproduce gene-memory through dreams. During these periods, one is subjected to the lower suggesting drives and wills of persons in his blood line who have long ceased to live in earth. From this, the world-dweller action or world-tempter can come into the life of the initiate.

The ancestral memory is seeking to immortalize itself at all times in its offspring or progeny. One must open the grace of ancestry and die to the ancestral forebearers' negative inclinations and compulsions. Each disciple and initiate undergoes these sifting sexual-initiations within the second chakra before he can unite directly with his Father in Heaven.

The teacher or guru teaches the initiate that he must establish his own individualistic relationship with our Father which art in Heaven, the true Father of all souls in this earth. Until each person honors his physical-earth father, mother and ancestry from whence he comes by making a direct union with our Father in Heaven, he will continue to be the victim of the darkened side of karma through ancestral compulsions; and the sins of his physical fathers will be his heritage until he has come to know the true Father in Heaven.

LOVE AND ANCESTRAL MEMORY CELLS

I dedicate to spend some time each day quietly in contemplation and in meditation.

When one loses the love-theme in the heart, he is desolate, and in time becomes a dissolute personality. One should dedicate each day to reunite his love-covenant with the Covenant of God. Through rhythmic contemplation and meditation, one maintains an orderly and perfected way of life.

The Angels, so close to God, live in an incessant state of jubilant joy. One who loves touches something of angelic joy. God is love; and where perfect love is, God is. Hunger for love is greater than hunger of the body.

When one dedicates to love God with all his heart, his mind and his soul, love will reorient him to a world of perfect relationships. The ancestral memory cells of the body will respond to the curative and exhilarating elixir of love.

If there be ancestral memory of non-loving and abuse, perfect love will cleanse away the mirrored offences registered and mirrored in each cell of blood-memory. Through perfect love, the barrier residing within Family-Atom association shall be overcome.

Love is a divine *filter*, infiltrating the darkened and fearful portions of the unknowing heart. The abuse of love through lust, covetousness and possessiveness divides the heart and the mind—and one lives in a world of doubt, uncertainty and insecurity.

When the disciple dedicates to love, he will come to see the transforming and miraculous processes of love. To love is to forgive. To love is to sacrifice. To love is to act with mercy, with charity. To love is to share. To love is to articulate. To love is to see, to discern, to behold. To love is to learn of the divine origins of all things and of all Beings.

SEVEN MEDIATIVE COVENANTS

1. My first covenant is chastity, for I would live with honor within my own self-esteem.

2. I shall make a covenant to surrender my lesser will to the Will of God, for I would know self-control.

3. I shall understand God's Covenant to man when I behold the wonders of the sky, the sea, the earth.

4. My heart is the body and encasement for my covenant to God.

5. In the beginning of this eternity, the Word or Covenant of Love was sent forth. Each word of pure love upon my lips is a sound of creation.

6. When I cherish and hold reverent what my eye beholds, this is the covenant expressing the Plan of God.

7. My first love shall be for God, and all other loves shall be of God.

ANCESTRAL ROSARY

> *The purpose of the following visualization exercise is to be free from the burden of one's own sins and the sins of his ancestors or "fathers."*

Once a week, preferably Saturday or Monday, one should repeat this exercise to cleanse the subconscious compulsions inherent within the automatic processes of gene-memory. This exercise is a confrontation with the genes on a cosmic level. It is automatic that once one steps out of gene-memory karmic tensions he makes union with the True Self, which is centered within the Great Atma or God.

The Ancestral Rosary exercise is to free one of the burdens of guilt saturating gene-memory. One moves out of his own karma and gains the higher gene-selecting powers, that he may reincarnate by choice through parents of high evolvement.

The Ancestral Rosary is to free the gene-negativity and to produce the quickening of the virtues within the Family-Atom. This Mantram produces miracles within the Increase of God. Families blessed with the use of this Mantram respond with unusual changes and quickenings in love for one another. The virtue-grace in the using of the Ancestral Rosary blesses the one who speaks it, aiding him to overcome self-hatred, separateness, competitiveness, jealousy.

> *1. In your thoughts, visualize a chart of your ancestral line, first placing the names of your father and mother. Then add the names of your grandfathers, grandmothers, great-grandfathers, great-grandmothers, and all other relatives that may come to your mind (sisters, brothers, aunts, uncles.)*
>
> *2. Speak these words to each ancestor separately in the following order: Father, mother, father's father, father's mother, mother's father, mother's mother.*
>
> ***I ask forgiveness of you and I forgive you. I pray for your prospering wherever you are. I pray for an easy birth and an easy death for you in all worlds.***

SAMSKARAS

> *Night instruction enables one to be initiated beyond gene-karma or the gene-samskara retention which inhibits a whole and complete spiritual dedication.*

The *samskaras* are the seed-tendencies one brings from past lives of past actions. These are more often thoughts than physical actions. The samskara confusions, irritations and tensions are more on the mental level than on the physical act level. When one is reaching for higher spirituality and the use of his mind as a spiritual vehicle, he is more likely to bring his samskara-action from the mental plane and from the subconscious, and thus his thought world is his work: the thought world and the understanding of what consciousness is; the understanding of what motivates consciousness and the understanding of how to flow with the consciousness; how to observe the consciousness, and also how to expand it.

Each person has an etheric energy-body which is mechanized by mind and mental forces accumulated through many lives of testing and trying. The samskara-colorings of good and evil are kept alive within these subtle etheric energy-processes.

It is rare when one enters the world with a healthy etheric body free of negative samskara compulsions. It is grace when the chakras have been purified through former-life spiritual initiations.

The lowest chakra at the base of the spine, when moving contraclockwise, prevents one from being aware of his spiritual potential. He expresses his instinctual nature through gross sensory, lustful cravings and sensuality.

In the post-Edenic states, or after man took on a coat of skin, the Undersoul was a subjective vessel having union with the inward laws in the etheric body. As man progressed in earth and became more egotistical and self-willed, the primal instinct in his purer sense was submerged subconsciously as samskara fire. This fire then became the substance for the Undersoul. In all creation, Undersoul fire is intelligible fire with sentience or sense recording.

Before one came to this world to serve the soul-need of his mind's expansion, he was aware of the need to accept the blessings of divine obligations and the disciplines accompanying the spiritual life. So it is that a true initiate never feels that he must blame others for that which has been blueprinted for him by the karmic impressions or samskaras dwelling similar to ravening lions, tigers or wild horses in the subconscious depths of the lower quelle portion of the mind.

The blessings of being aware of one's own Maha Guru or Teacher support all true initiates during trials, temptations and challenges. However, if one fails to observe the Laws, he regresses as an initiate.

The Master Teacher or Guru enables one as an initiate or disciple to live within the causal Buddhi mind. In this, the initiate is mentally strong and alert during accelerated trials or initiation. The samskaras are controlled, balanced and timed to the flow of the outward life.

The higher unconscious portion of the soul is a bridge of light between earth sentience and the spiritual superconscious. The Guru Master-Teacher teaches directly to the higher unconscious or higher quelle so that the student or disciple may erase the samskara, darkened, earth-impressionable remembrances dwelling within the lower subconscious or quelle.

Confession is the door to self-honesty. In the lives of some who reach toward the Path, the time will arrive to go through the door of confession. However, if their sense of shame and guilt is unbalanced, they will turn back or fall back. The overemphasized and heavier weight of their burdens will lead them to that destination where they can work out their unfinished karmas. If one has a Living Teacher in this time and has given service, respect and trust to the Teacher, the Teacher will take him by the hand and guide him over the abyss of his samskara darkness.

The Old Testament is filled with dual-samskara initiatory trials which are a part of prototypal existence when one is yet karmically bound to the downward ancestral flow of duality inheritance.

The samskaras in the subconscious are fed by the lower devachan duality suggestibles. Through Christ, the initiate makes a victory over the duality devachan entitized suggestions.

Since the coming of the Christ, Kundalini power in the

Western initiate has drawn upon the individual subconscious memory of former more subjective states in Kundalini function. Thus, the Western initiate seeking to open the Kundalini power and attain superconsciousness by opening the higher chakras is compelled to undergo the stirring and revival of the vigorous samskara genetic images of former-life sexual compulsions. The initiate of the West must master and conquer any over-aggressive sexual tendencies of the over-forceful self.

Jesus was sent in Judas an artist with the artistry of the black master. Satan uses those having the subtle, insinuative, innuendo cultures. These innuendo cultures are satanic samskara-tendencies which are used by the arts of inquisition: to torture, to humiliate, and to distort by lies and insinuations the truth of the prophets.

The torso portion of the etheric Tabernacle, Aaron's Lodge, where dwell the three basic Initiations—Hierarchy, Will-Power and Surrender—works with the Action Principle. When these are stirred within the subconscious, they send forth the samskara-tendencies of karma. One comes into direct contact with his own motives. One must also begin to intuit his past-lives' history. He opens aspects of his memory which are valuable to him as a cargo he takes to the Holy of Holies for definition and for guidance.

War is caused by the collective subconscious of man. The samskaras which man has not fulfilled become guilt in him and then he begins to act in a rajasic manner and has a collective desire to kill. This comes on men en masse. That is why when a war starts men cannot stop it, because it is a flow of collective consciousness of the rajasic guilt of men which must in some manner make restitution through sacrifice of blood. Those who are killed in wars make sacrifice on martyrdom levels. If one is killed three times, he does not have to be sacrificed in war again.

The most obsessive form of fantasy is the inheritance

from one's ancestry. The most deluding characteristic inherited from the impressionability of karmic inheritance through the ancestral line is that one is impressed to do the negative things his ancestors have done—and thus thinks himself to be doing the right thing, such as, the taking of lands and possessions belonging to others due to the coveting nature within one's ancestral stream. This fantasy is passed down subjectively to those who have the samskara tendency within their own natures. This is the cause of wars and the taking of lands.

The Destroying Archetypes are regulated to destroy through the Time-system in this eternity system. Man's Time-mechanism is presently being overstimulated by the Luciferic or Satanic Archangel, inspiring people to sit at the feet of the media, or TV, and to be held hypnotically into the dramas of violence. These dramas implant samskara-seeds of evil suggestions and motivations into the receptivity of the present or modern aspect of the subconscious of the individual and of the world.

The Commandment *Honor thy father and thy mother*, which relates to the Family-Atom, is especially activated by one's seeking to overcome the samskara-tendency of sins inherited from his ancestors. These sin-memory processes lurk within the subconscious or appear as samskaras or tendencies to sin in the impressionable suggestions inherited from one's ancestors.

> *The daughter inherits from the father the samskara diseases. The son inherits from the mother the samskara diseases.*

Detachment

> *Detachment is the ultimate goal and gift of purification.*

In some systems of psychology, there is a revolt against the Ten Commandments. The more one keeps the Ethic and the Commandments, the more he is free from tyranny. The only time one has total detachment is when he is full of Ethic and full of the Commandments.

Logic is the Logos of the mind. Logic is the higher rationale through which man links himself to Creation. Attentiveness and concentration-exercises build logic. The closer one comes to the higher logic, the more he must detach from the heated aspects of the lower mind.

The mastery of sin indented in the Undersoul gives detachment. On mastering detachment, a third aspect is born; one moves out of duality away from sin. Conscience then is free to act as a chemical catalyst to devour sin.

The will beyond sense is non-possessive, desireless. To live in the Edenic morality aeonic-grace, one must release tension-laden desires to the Will of God. If one is free of tense-desire willing, he has freedom, liberation and salvation. One attains selfless desiring by selfless service, detachment, renunciation, sacrifice, giving and trust in God as the Will and the Plan.

Anyone who enters the Path with the intent of healing others and teaching others enters into Intentional Suffering. Intentional Suffering is an empathy far deeper than sympathy. It is an empathy carrying both divine responsibility and detachment.

The secret of detachment is non-acceptance of reward.

Over-self-importance is a form of ego-madness founded upon an oversurplus of pride. In the Heart Chakra, one is taught the law of detachment. This law manifests first as being able to detach oneself from his own sense of self-importance.

The Anahata Chakra (Heart Center) has twelve petals. This lotus is spoken of as the lotus of the unstruck sound, or where one can know the Transcendental *Esse* or Tanmatra etheric Essence of all existence and knowledge, all bliss. Its element is *air*; its sense is touching; its pranic vital function is *love, detachment, de-manifestation, releasing importance.*

In the lowest chakra or Muladhara Chakra, one is initiated by Shakti through Maya, or the energies of Prakriti or Supreme Nature. In the Svadhishthana or second chakra, one must master self-deception and delusion to produce the power to will. In the Manipura Chakra or third chakra, one must master illusion through the use of detachment from desire. The initiate must overcome dependency upon the fantasy-chimeras of the astral nature. In the Solar-Plexus Chakra, one masters the Law of Archetypal Symbology, and takes command of the forces within ether, prana and akash. He thus transposes the gamma rays of the Sun and the ultraviolet sun-rays of light into the cosmic white-light, using the Esse or Essences of the white-light to extract from ether the higher pranas and akash, that he may heal and cleanse.

Each imaged portion of God producing the real of Self is a vital aspect of God. In one's initiations of attachment and detachment, he must learn the difference between self-importance and self-worth. To do this, he must stand off from himself, becoming the Witness to his motives, desires and true intent.

He who hath acquired the peace of true detachment hath entered into the golden Light of his etheric mold and naught of the outer world may touch or claim. He who hath stayed in the Great Silence within the night hours of his sleep hath acquired over the many periods of discipline the serenity and poise which cometh of detachment.

Attachment pertaineth to the possession or claiming of the soul's right of those nigh and those associated in karma. Attachment pertaineth to things of material as of supply which possesseth and mastereth him who owneth it or claimeth it. That which pertaineth to attachment pertaineth even to country, nation, and even to the blind vision on which man holdeth his God.

Detachment to the spiritual disciple cometh from the many strivings, the many climbings, and the many non-climbings. That is detachment which giveth over to the Whole; and in the giving receiveth of the All.

The spiritual disciple findeth the world a difficult place to assert himself without aggression, to stand in that which he believeth without rigidity, to live that which he teacheth and knoweth without criticism, to build a boundary without a barrier, to build for himself a life in which the most pure might come without sullying it from outside interests, and to so remain poised within the great system within Light and Love as a disciple. He findeth often the word detachment *interpreted by those near him as indifference or of selfishness, but he becometh of the silent, and standeth in what he knoweth. And portion by portion in Light, Time, the great healer and minister of the Christ, proveth all things in which he standeth and for what he standeth.*

He who hath the courage to take over detachment buildeth his life quickly above the stream of karma and hasteneth to slough off that which interfereth in the point of attachment. He blendeth attachment to Love much, but to claim not. He blendeth Love, the non-claiming Love, with Light, and on this all come into their own in the precious relationships which come out of the blending and sharing in Light.

Part Two

The Three Heavens

THREE HEAVENS	ARCHETYPAL KINGDOM KINGDOM OF GOD REALM OF LIGHT	THIRD HEAVEN
	Vibrancy Barrier	
	SPHERES OF LIGHT OR HALL OF WISDOM	SECOND HEAVEN
	HIGHER ASTRAL WORLD ISLANDS OF LIGHT	FIRST HEAVEN
	Abyss	
MAYA	LOWER ASTRAL WORLD	
	PHYSICAL WORLD	

THE THREE HEAVENS

10

THE LOWER ASTRAL WORLD

An initiate of the spiritual life commands, understands, and has access to both the astral and spiritual worlds. An initiate is in command of his own ego, fully aware and conscious, inwardly and outwardly, of all encounters astrally and spiritually.

The Astral World

The higher astral world is the First Heaven. The lower astral world relates to man's subconscious mind in death and in life.

The word *astral* means starry plasma; star or planetary reflections. The energies of the planets are reflected into the astral world. These energies influence the actions of men. The astral is a unique, unceasing, fermenting, mirroring and moving action, producing in man a state of emotion, mood and inductive feeling and thinking.

In the Bible the astral world is called the *serpent* or the *dragon*. The term serpent relates to the incessant movement of this serpentine, heavy, starry-laden ether. The lower astral world in the interior of the earth is stimulated and kept alive by the coil of the magnetic belt around the earth. The lower astral world is the dark side of the hidden deep. The lower energies of the planets fall into this serpentine coil—hence the name *astral* or *starry*.

The reflective portion of the lower astral world forms the sin-body of the earth. The energy-laden portion of the astral world is a gelatine-like substance sustaining resiliency and movement in certain etheric forms and shapes.

Pressure from this lower astral coil plays upon the lesser etheric body of man and also upon his sentient emotional atoms. The negative mental powers of man respond to the agitation of the lower astral world. All preparatory initiations are activated first through the lower astral pressures. Man must master this astral serpentine sea, that he may enlarge his sphere of action in the world and in Heaven. All initiates first are confronted with the astral sea and undergo the great abyss or reflective sin-body initiations to enter Heaven.

"Now the Lord had prepared a great fish to swallow up Jonah. And Jonah was in the belly of the fish three days and three nights." (Jonah 1:17) The story of Jonah in the Bible is a story of an initiate being caught for three days in the whale or the "great fish"—the astral sea. When one is ready to overcome the astral sea, he is literally swallowed by the serpent or astral sea; he experiences the sin-body initiations, and he undergoes a magnification of every possible error in his emotions and mind.

The satanic forces use the lower astral agitated ethers to transpose pictures of enticement upon the minds of men. The Bible speaks of the time when the tumultuous sea of the lower astral world will become as a "sea of glass." (Revelation 15:22)

Science has discovered that the magnetic belt around the earth undergoes potency variations and periodic reversals of currents. Spiritually, when men experience the diminishing of strength in the magnetic field surrounding the earth, this results in a lessening of the satanic powers, and permits a spiritual mutation to take place in the chromosomes and genes of man, thereby enabling him to extend his emotional and mental atoms and receive more directly from Heaven.

Biologically, some earth creatures, plants, and animals during these times are withdrawn from the earth, never to be reproduced again. God has so timed His creative tides in the world that man and Nature must be given healing respites or interims during which man may come closer to Heaven. However, for those who aim for unceasing communion with Heaven, the lower astral world is a sea to be crossed. The story of all martyrdom, all saintly acts, all illumination, is a story of the soul tossed upon the tempest of the astral sea.

Unknowing men live in the world subjected to the astral tides, caught into nebulous traditional interpretation of their soul-life. Such men are victims of the lower cosmic forces and suffer more intensely the anguish of human trials. Their only relief is in the world of dreams during sleep, where their souls provide a freedom to observe and record a side of Heaven unknown to their physical minds.

Jesus is a walker upon the waves of the astral sea. His Presence in Heaven enables the spiritual adept to move over the writhing abysses of cosmic tumult. All spiritual adepts who come under the mediation of Jesus are free to go "in and out" with angelic protection.

> *If I have told you earthly things, and ye believe not, how shall ye believe, if I tell you of heavenly things?*
>
> —*St. John 3:12*

As physical light encompasses and permeates the earth and the world of man, the astral essence permeates and encompasses all feeling in the world of man. Man commands the astral world by the use of his higher will and higher mind.

Each of the seven major glands of the body contains an astral counterpart or starry planetary center which is stimulated by the lower and higher vibrations of seven planets: Saturn, Jupiter, Venus, Mercury, Mars, Uranus, Neptune. The vibrations from these planets shape and subtly direct the individual ego, re-enforcing its expression.

The reflection from the planet's light makes the astral world. If there were no Moon or Sun, or Mars, Mercury, Saturn, etc., there would be only astral latency, not action.

When any spiritual work is done, it is done not in the astral world, but because of the astral world. The astral world is merely the pillar supporting the Spiritual Archetypal Formulas and Forms.

In the Scientific Age, the spiritually-inclined persons with mystical aptitudes will be less endangered by psychical pitfalls than they are at the present time. Within the last century, the untrained mystical mind was exposed to the lower regions of the astral world. The Scientific Age now coming forth contains a higher note of spiritual science, which will enable all sincere disciples to fulfill certain ethical techniques, and thereby assure themselves of the certainties of the Higher Worlds and of their aid to man.

EVALUATION TRIALS

The Lion-Initiate is one who has mastered the forty-nine initiations of the astral planes. The Eagle-Initiate is one who has free entry into the Spheres of Light. The ancient symbol of the Winged Lion pertains to the adept who has the power to interpret to others his knowledge of the Higher Worlds.

Within the astral world are seven planes; and within each astral plane are seven regions. These forty-nine regions enable a person to undergo the initiatory trials of evaluation.

The seven astral planes are: (1) grotesque; (2) fantasy; (3) wish; (4) the plane of reflected records, that is, the archaic records of the earth, the reincarnation records of the masses and civilizations, the embodiment records of the individual; (5) the initiatory; (6) the prophetic; (7) the spiritual.

The two higher planes of the astral world are inhabited by the Saints, the Bodhisattvas and Buddha. The two higher astral planes are referred to as "Nirvana" in the Brahminic Archetypes.

Beyond the astral world, in a higher degree of Light, are the Spheres of Light, the Realm of Light, and the Archetypal Kingdom or Kingdom of God.

All persons who would serve as mediators for the Invisible Worlds must first undergo the forty-nine self-evaluation initiations. These initiatory trials consist of the Moon's reflected-sphere initiations pertaining to the three lower planes of the astral world: grotesque, fantasy, wish.

The second phase of the self-evaluation trials relates to Saturn's reflected sphere, where the initiate works with Sanat Kumara. Here he may read the archaic records of mankind's history and trials in earth. The third phase of the self-evaluation trials relates to the higher astral world, where dwell the Saints and Buddha.

When one is being initiated for the Saturn's reflected sphere, he must cross the abyss between the third and fourth astral planes. This is a ring of darkness; and, because he has yet to qualify for unrestrained night-flight, he must be assisted over this abyss by the Angels.

In the ancient days, the sages of the earth, knowing the workings of the astral world, were given number-formulas or a kabalistic system of ritual to enable them to overcome their lesser emotions. In this age, those who seek and dedicate to the Light are given a system of *declension* enabling them to use mantramic formulas through which the combining of words enables them to rise above the astral planes and to be initiated into the Spheres of Light, and thus become aware of the greater powers of Mediation.

> *The earth may be likened to a cup. The astral world and the life in it are the handles of the cup. The Overdirecting Intelligence, or the Father, holds the cup. That which is in the cup is mind.*

The higher astral planes are to the Spiritual World what air is to the earth world. The ethers in the higher astral planes make it possible for men to mediate between Spirit and matter. Since the higher astral planes are the meeting places between Soul and Spirit, they are the means by which Soul carries out the formulas between body and Spirit. Jacob's ladder in the Bible is an illustration of the actual experience of penetrating the higher astral planes to reach the Spiritual Realms beyond.

To avoid the world *astral* is to not receive the full value of its ways and works. To say it is the place of "desire" takes in only certain actions.

The Teacher, having mastered the gravity planetary initiations, understands the Laws of Energy as reflected in the emotions of the disciple. He comes to still the sentient atoms in the astral tumultuous trials of the disciple. Thus, he works in day as well as in the night-initiatory spectacles of Maya, or mirage obscuring-fears, to give calm security, to give support, trust, and to give reviving strengths to his disciple.

The Living Teacher works to assist his disciple in the control and dissolving of his present-life karma. One is in a state of astral confusion until he acknowledges with gratitude the ones who can show him the way and the why of the soul.

Until one has command of the three lower astral planes, he is not ready to be a Teacher. The unmanifested processes must flow through him and be made manifest.

Temptations abound in the world of man and in the lower astral worlds. Pleasure-principle obsession leads one to the temptation trials.

During the materialistic periods of initiation, the Western closed-fist teaching, withholding the keys to the great Dharma, restricted the lesser mystery schools—the full Dharma being by-passed by astral light.

To be snared into an astral net of a false guru or teacher, one has in some former life abused psychic laws. From such teachers one receives disillusionment and sometimes can be led into delaying instructions.

False teachers teaching the glamor side of the psychic life are to be avoided. False teachers desiring their own egos and their students' egos to be magnified by adulation are products of astral light rather than spiritual light.

Fortunate is the one who knows his Teacher to be a trust fortress. Such Teachers come, using the divine-intuition, and knowing when and how to make equal testing and protection during initiation.

> *Send gratitude and blessing to the Teacher or Guru-line each day, and thus avoid the pitfalls of deviation. Recognize Jesus as the Teacher over all Teachers.*

THE GREAT ULTIMATES

Pluto blendeth the lower musical sounds of the astral underworld onto the higher note through the coming day and coming Light. Men reveal to themselves as never before the stirring of the deeper mind into the subterranean passages of the past and find their freedom in the freeing of the Center Q in the base of the skull, corresponding to the number 8, the planet Pluto, to the great Spheres above the astral, to the overcoming and mastery of the dark.

That which soundeth the tone of the 8 on man freeth his soul-light above tribal memories as of ancestors in karma. He blendeth his soul-tone onto the Music of the Spheres, and in so doing aideth in that which buildeth the planet Earth in its lifting to its Heaven-Tone in God.

Music received from the lower astral world playing upon the senses is given to man that he might express himself through movement of body.

The disciple, working with the abstract light over the city, sees how the Hierarchs, the Great Ones, and the Angels use the Music of the Spheres to inspire the builders of the city. He sees that the abstract lines in coming buildings in the physical world are held together in the fifth plane of the astral world by tones of music. When the physical building has fulfilled its purpose on earth, the tones surrounding the abstract form cease their sounding in the fifth plane of the astral world.

Men concerned with their own karma have become selfish in that they see not the karma of another. The why of karma is to be revealed and tenderly received in reflecting Light. The mirrors which have obscured man to man through astral light now open in the sixth plane reflecting the astral world and bring man to the vision as of one in Christ.

Infection in the Earth comes from the astral or feeling world. That which is contagious comes from men on similar rays in which sickness pertains to an assimilation or similarity between ethers in men.

As long as there is separateness in the world, sickness will be a means of lowering man's vibration that he might release his egotism and lower self to reach the Higher Self. As long as men understand not the blending within bodies, there will be sickness of curious and malignant nature possessing them. As long as there are jealousies, infections will turn into degrees of cancer and malignancies incurable to men. Infection is that which turns not itself out in soul-light, but turns back on man.

The astral world is filled with planes of correspondences. Bacteria on earth has an exact corresponding correlative in the astral life as thought. When great epidemics occur on earth, it is because mass thought has become out of order.

Nothing physical ever completely heals the physical body. Healing first takes place in the mind, working through thought correspondences by faith. If the thought causing sickness is etched in from other lives, this sometimes requires many lives to set it aright. A so-called incurable disease is a cell imbalanced by a bacterial agent out of tune with its thought correspondence in the astral world.

All men are subjected to change. The Ultimates exist in the World of God. When man discovers and relates himself to the seven Ultimates, he is then working in harmony with God, and the result is good.

The physical world with its constant state of change exercises the soul-powers into faculties for the use of the world.

The saying of the Lord Jesus to "be of good cheer" is the outer cardinal expectancy of one spiritual Ultimate. The Lord Jesus gave the seven Great Ultimates. When the Light of the Christ shines into the earth body, there is a great responding in all the planes from earth to Heaven.

Even as in the physical, there are no vacuums in the astral world; however, the reaction to things is quicker in the astral than in the physical world. When one thing is removed, another immediately takes its place, always according to the order of what it replaces—its part or its replica. It is in the power of thought to replace an inferior thought with a superior thought in the range of consciousness.

A cheering mind, a cheerful thought, keeps the level of creation on the level of the Brothers' work in Christ. This is sustained by unselfish thoughts. Unselfish thoughts work with the Ultimates, which are instantaneous in their active principles.

> *To be a part of combined soul-light with spiritually-inclined souls; to be called to do a spiritual work—is to expose oneself to Dweller opposition forces in the physical and the astral worlds. Adverse trials invariably accompany every phase of initiation to the spiritual life. Through such trials, the weak are sifted; the strong are cleansed and chosen to stand, to teach, to heal. The strong do not fall or fail in their mission, that they may become transmitters of the Light.*

THE LOWER ASTRAL WORLD

> *That which affecteth the disciple in the lower astral planes cometh for experience and strength and courage. If he overcometh one astral beast in the lower astral world, he hath overcome 20,000 men who, as enemies, rise against him in thought.*

Gravity in a third gradation is beyond Earth-gravity limitation. There is no other way one can explore the Spiritual Worlds until he learns how to move beyond gravity on the physical and astral states. Beyond the astral gravities, there are Godly energies that are the impulses which command and give maintenance to gravity.

In the lower astral world, the great Sound Current is toned down by the gravity reaches of the earth and of the lower vibrational rates or subelectromagnetisms. One expressing the lower side of his ego undergoes a frightening experience in the lower astral world.

All higher initiates are at some time in their lives exposed to the lower psychic forces as expressed by subtle beings and human beings, that they may better understand the psychically-charged atmosphere produced within the lower astral planes, and also that they may have command and mastery over any reflected soil of the psychical nature within their own natures as retained in their shadowed bodies from former lives.

The most important aspect in initiation of the ego is will. All initiates to be used of God must come under the Will of God. When any person intrudes upon the will of another—through uninvited hypnosis, through subjective astral mesmeric influence, or in any degree seeks to divert the will of another person and to exploit it, to gain or to in any manner attain a certain power over the mind, emotions and will of another—this is of the dark. All initiates understand that the true life of willing is also the life of being. When one on the path of the spiritual life surrenders his soul-life will to any human person, or subjects himself to the will of another, he invites the shadowed and lower psychic world; and, in the processes of initiation, he will invite the forces of the dark who will challenge him and entice him into diverting chimera-like situations issuing out of the astral world.

The lower astral world is a world of tumultuous forces. One exposes himself to the lower-astral-world tumults when he has within himself any forceful desire to attain psychic powers and to thereby gain influence over the lives, minds and affairs of others. To truly experience the finer side of the play of the psychic forces and discrimination concerning psychic energies, psychic forces and psychic powers, one must be willing to serve God and to earn these powers.

Man presently is in a delirium of astral obnoxinity. He has dared to profane his atmosphere given to him as a green pasture of stewardship.

Insensitive exaggeration accompanying raucous astral music; the use of drugs; sexual experimentation producing perversions—all are influenced subjectively by the dissonant sound currents in the lower astral world which has for ages reflected a form of condensed or collective repression in man's desire nature. His nongiving; his lack of interest in learning and knowing; his greed at the expense of others—these have accumulated as energy tensions in the lower astral world and are now playing upon the *Undersoul* of man. This condensed sound in the lower astral planes is producing a theme of music obscenely heard upon the suggestible sexual nature of man.

Mankind, overlong too gross and dependent only upon its materialistic existence, is now being flooded with astral phenomena incited by astral forces and fantasies. When this has played out its force, and the last laggard souls are finally moved into this world or earth existence upon the conveyor belts of reincarnation, the stage will be set for a religiosity return to God through worshipful processes of beauty, culture and creation.

He who lives within his emotional negations sends out a negative strain of energy and power onto the lower regions of the astral planes, which in turn is used in that which breaks down and destroys in the law of disintegration.

God, as Creator, the Christ, the Son, and the Holy Spirit, the Activity, carry within the astral world one note—a note of resolving that which adheres, clings and crystallizes.

Men's negations in their accumulations contribute to the

power within the astral world, which in turn, by rhythms come back to men, falling upon them and reflecting the World of Tone as sound on earth. The turbulent waves and rhythms of sound from the astral world exhaust and deplete men.

Men come into the earth to gain experience, and on receiving experience record it in the soul-light. On the rest after a day of life in earth, the soul's light is gathered around them as the soul-garment. It is this day which is recorded in the starry light of the world-soul and of the planetary rays affecting the astral world.

Through materiality and through partial knowing, men attract the unreal. In their untrained desires and need, they seek teachers who give them of the glamor, the false and the partial. Glamor is the first step on the left hand path, where men see reflected in contortions of their own desirings the patterns in the pictures of the twilight in the astral world.

Those having received the partial truths of the lower astral world have reached a state in earth in which—through the force of pressure, war, loss, lack of understanding between God and worship, lack of understanding between life and death—their minds express a psychic world.

The experience within the astral world brings pictures. The Masters of Light work constantly that these pictures focus the True and place themselves into the mirror of men's souls, that their minds' light might receive them accurately in accordance with the Tone of Truth in the Spiritual Worlds.

Facing men in the next 79-year pattern is the building of an expanded consciousness. The true meaning of the astral world, its work on men's subconscious fantasies, its work on men's false desirings, shall be understood. The germ of an idea now awaits to come into the Earth. It comes through the Master's Light and those working through the Light of the Hierarch. This germ to come through a human being in Earth is to place a simplicity in the tone of all religions, and the tone as an atom of mediation is to blend the many religions in their scattered creeds.

True worship takes of itself in the next 100-year cycle a renewal after the Light in God through the Sonship in Christ. That which comes to man comes in a vessel of human form which brings the Light in Earth and reneweth the Spirit of God as living in men between Heaven and Earth as one.

> *That which affecteth men in the writing world not pertaining to Truth, or that which is of fiction occurring in the writing of men as of the imaginary on which men bring in the profane, the vulgar, the disrupting ideas out of the lower astral regions, createth karma on him who writeth and on him who readeth.*

I. *Grotesque.* The first plane of the lower astral or duality-devachan world is the plane of withdrawn etheric molds of life once existing on earth. This plane also contains the negative thought molds of men, and pertains to the subverted subconscious, where one is exposed to the ancient, negative records of past lives.

The grotesque level of the astral world reflects the sin-body of the earth. The four lower planes of the astral world are the recipients of the lower vibrations and energies of the planetary light. This sub-planetary energy produces a chimera, mirage-like effect upon the emotions of man in life and in death, producing glamor and mesmeric effects upon the lower mind and senses of man.

The Devachan Kingdom exists in the First Heaven as well as in the lower astral world. In the lower astral devachan planes, the Dual-Devas work with the Cherubim Angels upon the lower chakras to develop man's psychological nature, and thereby produce his birthright of individual creativity.

Delight in the pleasure principle is what stands between a person and a true relationship with God. One often thinks he is having a bliss-experience when he is having a sensual pleasure-principle experience. Sensual pleasure-principle experience occurs in the lower levels of the devachan astral world.

The grotesque astral-devachan gives a bloated, misshapen body. Much yang food is needed in nutrition for these persons. During cyclic glandular periods, one must be alerted that he does not expose himself through pleasure-principle desiring to this grotesque astral devachan world, where self-indulgence becomes the enemy of the body, the emotions and the mind.

The Kali devachan woman works through the grotesque, wish and sexual-fantasy levels of the astral world. Kali devachan woman, when sending from the grotesque level of the astral, is large-boned, fat, obese, sensuous.

An indulgent posture in prayer sets up a barrier to pure alignment and only the atavistic impulses of the astral world are reached. The solar plexus rather than the heart is activated, and the actions of the senses contest the pure entry into prayer.

The human race has received the expression of sex so as to propagate the earth. In cases of multiplicity propagation, as in the present time, the astral body of the earth is over-extended, exposing one to his own lust and immoral nature. It is the duty of all who are pure or seeking to be pure to teach the laws governing sexual life. This can be done only through the angelic wisdoms supporting birth and death.

The Personal Angels of Pure Desiring, Guardian Angels and Recording Angels are Guardians against astral pollution.

II. *Fantasy and Purgatory.* The second plane of the lower astral world is the plane of subtlety and psychical trials. If one has retained psychical powers from former lives, he is exposed to the black magicians' telepathies of this psychical plane and to the entity shells of the dead during sleep and also in daytime thoughts and feelings. Through self-mastery, one insulates himself from the psychical trials and qualifies for the higher or greater experience in the First Heaven.

That which reflects the dark passions in the lesser regions of the second astral plane gives birth to poisonous plants in earth. Poisonous plants in the physical world relieve the astral world of the force within tension; for men's poisonous emotions—such as hate, force, greed, pressure, selfishness and one-sidedness—are absorbed into the poisonous plant. And from the plant comes the remedy which heals man.

Each poisonous plant and herb of the earth corresponds to a hate or weakness inscribed upon the lesser regions of the second plane of the astral world. That which is between the poisonous plant and the astral world as growth relieves man of his hate, and aids him in the bearing of temporary karma. But that which pertains to man's karma as to his being hated, entering into his thought world, goes into the tree and brings Angelic mediative helps of healing within the Spiritual Spheres. Man begins his work as to his karma when hate enters his thought world. When hate enters only his emotional world, the plant aids and relieves him. But when hate enters the mind, it becomes malice and of the dark, and then the tree bears it. Even as the cross bore the Christ, so does the tree bear the thoughts with the aid of the Angels.

> *The subtle pressures in the astral world come to the time of explosion on men's minds and bring the condition which purgeth the world and beginneth the cleansing of the barriers between race and tribe, nation and continent.*

III. *Wish, Bliss and Paradise.* The third plane of the lower astral world acts as a mirror of one's deep-seated desires experienced in daytime and in dreams. On this plane, one qualifies for personality placement.

The planet Neptune is a galaxy mediator, taking from the great galaxy system supporting the solar system of the earth. Neptune is an initiatory planet. One being initiated under Neptune is exposed to the fanciful chimeras of the wish level of the astral world. If one is untrained, the central nervous system is affected, producing hallucinatory states of consciousness. A mature person undergoing Neptunian initiation frees the vagus fire in the nervous system. The pineal crown produces the power of revelation. The highest aspect of clairvoyance is produced through Neptune's osorius or highest pranic atmosphere, that is, wave lengths of revelation superior to mortal sense.

> *The highest region of the third astral plane is a reflector for Nature's symphony.*

"And Jesus said unto him, Verily I say unto thee, To day shalt thou be with me in paradise." (St. Luke 23:43) Paradise is a peninsula or bar of heavenly light located between the purgatorial caverns and the First Heaven. All persons touch paradise after death. The wicked, being asleep, are unaware of paradise; the good are awake. Paradise is given of God, that all who die might experience for a brief time a preview of Heaven. In paradise, there is beauty such as men have never seen on earth. There is incessant action, order, rightness, and no resistance. Paradise is every man's plateau—a pause after death to give one stamina for the next phase of progression following death.

Paradise is similar to a sandbar at sea. It is a springboard to either the darkened caverns of purgatory or to the lighted corridors of Heaven. On each side of paradise are whirling, intelligible currents which draw one to his rightful place. The Guardian Angel, working with the record of the soul, determines whether one crosses over the bar of paradise into purgatory or into the First Heaven where dwell the Risen Dead.

> *The East is subtle. It is subtle out of the subtle regions of the astral world. It representeth the lower acting regions within the third plane of the astral world. The East or the Orient, having lived through Ritual of the astral, now moldeth her patterns after a mystic mystery, and men of the East move by impulse not known to themselves, and act upon life as a force in the world.*

IV. *Archive Memories and Embodiment Records.* The fourth plane of the astral world is the beginning of the First Heaven.

The four lower planes of the astral world act as a mirror. The fourth plane of the astral world mirrors the Greater Archetypes. These become the etheric braille reflecting the records of the past aeons and ages. The etheric braille also shows in part the things to come.

The Master's Light overshadows man as akasia. In this, man touches the fourth plane of the astral world, using the fifth, sixth, and aspiring to the seventh power in Light. He creates energy out of the source of Light, and with the use of the will, brings the seed-atom of the heart to the point on which energy renews itself and brings forth life-substance, eternal, immortal, in man.

In the higher regions of the fourth plane of the astral world may be heard the Divine Cosmic Music. It is on this plane that celestial harmonies are heard by great musicians —and through their creation, these harmonies later fall into the outer ear of man.

If it were not for the octaves of sound within the fourth plane of the astral world, and the Islands of Light built by the prayers of men, the greater etheric Beings could not approach the earth.

The Animal Kingdom is protected by the Cherubim Shepherds of the first region of the fourth astral plane, the Species Angels and the Fauna Angels. The Cherubim Shepherds of animals enable animals to have a sense of humor; they direct animals to their true masters, and enable animals to contribute to the humor of man. The Species Angels overdirect the propagation of animals, the health and the vitality of animals, and the instinct of animals. If one wishes to heal an animal, he should call upon the Species Angels. The Fauna Angels work with all undomesticated animals. The Fauna Angels determine what time or age certain wild animals appear on the Earth and also when such animals are to become extinct or withdrawn.

> *The Aquarian Age of which men have spoken since the 1800's is not the Age of Peace, but is even as the Lord Christ said in the body of Jesus, "I came not to send peace, but a sword." So is that which affecteth Uranus and the Aquarian Age, bringing a tumult reflected in the astral world, affected in the deeper layers of the Earth and in man's outer consciousness as man against man in a spirit of competition inverted in the Earth.*

ETHER AND THE ASTRAL WORLD

The astral world encircles and interpenetrates every part of the physical world. Particularly is its action parallel to the central core of the earth, where it gathers a momentum to work with the etheric substance required for a more pliant action in astral-form pageantry. If it were not for this work, men would not receive symbols in their dreams and in their visions.

Ether is a coagulating agent in the astral world, even as it is in the blood of man—for ether is the mingling fiat. Ether is chameleon-like, in that the quality of ether is determined by that which it adheres to, lives in, and of. Ether has the power to transform the coarsest matter into the finest waves of energy within light. It is the means of retaining the living substance of Spirit on any plane—from matter to Spirit. It is the substance making possible the intermingling of all substances; and action within all realms, spheres, planes—objective and subjective.

If it were not for the ether accompaniment to the working of the astral world, that which is astral would result in a frozen or static substance; the astral world would be an enlarged abnormality attached to our earth world. Ether keeps movement, clarity, and geometric precision exact. Ether coordinates that which is supple in all planes. The study of ether's work will extend the range of symbolic patterns, enabling them to become an actuality to the rational mind. When one understands the transforming pliability of ether, and relates himself to the law of cause and effect, he can accomplish miracles as to precision and timing.

When ritual is recollected by the soul-power, and is orally used, ether becomes a directing and controlling agent in the astral world. Ritual uses ether to unfold the formulas supporting the functions of the soul.

In the world of earth and materiality, habits good or bad respond to the function of ritual used by the higher mind. Habits built by repetition unconsciously use ether to create the coarser manifestations within the world of outer form.

The astral world is sustained by the spherical movement of the ethers within the astral world, in which there are planes reflecting the convolutions of man's feelings. These convolutions are generated by the spherical particles within his emotions, moving in orb with the planets. Man's emotional process goes always toward the encircling of an idea, and returning it to its original intent within feeling. Thought originates in feeling, in desire. When thought has been clarified and acted upon, it returns to feeling.

Karma works in a spherical movement and orbit. When

one thinks with the lesser mind, this begins a geometrical pattern which is reflected in the astral world. Negative thought produces an angular pattern which is magnified by the spherical action in the astral world. This irritates man, and creates pain in the emotions and the mind.

The lower planes of the astral world are tumultuous mirrors, reflecting the negative emotions and thoughts of man. These are enlarged and magnified by the spherical movement within the lower astral planes. Thus, when men think negatively, it becomes an emotional contagion in the world, seeping first into the feelings and then resulting in the destructive ideas of one toward another.

The purging in great cities creates an impassable boundary in the higher regions of the astral world and a convulsion in the interior and anterior astral regions within the earth. One who reads the reflections in the interior regions of the astral world sees them reflected in the discards of the anterior regions within the earth. He finds a conflict between the two, a raging torrent between man's passions and his faulty aspirings, in which man climbs not for God, but climbs only for self.

The shadowy elementals dwelling within the sub-levels of the astral world work with men's cunning minds in cities and sustain the evil works of the dark. The Warrior Angels come forth against these shadowy elementals to cleanse the ethers of the city. There is a continuing war between the shadowy elementals and the Warrior Angels when a city has fallen into corruption. The shadowy elementals become bound under and the corruption is cleansed. The people in the city rise up to fight corruption; and evil men are cast out of the city.

The Astral World and the Afterlife

Ritual of today hath become a formula after the pride in men. Such ritual buildeth a body within

> *the lower astral regions, and maketh a sameness, a static condition; and men receive their sorrows, their conscience, their repentance, their griefs, their dead, their loss, reflected back to them after the agony of their own souls. Ritual ceaseth to give peace, but continueth as a memory and reminder of man's past actions, errors and sins.*

The law of gravity in the physical world works upon the physical body and the lesser etheric body. In the astral world, the law of gravity works in the thought. The gravity in the astral world, working with the Sun and the Moon, forms a cleansing matrix. All are drawn into this matrix after death so that the pain shell (lesser emotional body) may be cast aside. Sensation cannot reach into the higher levels of the astral light. Pain is the highest expression of sensation, contrary to the generally accepted idea that joy is the highest.

What is really joyous in the afterlife works through the highest astral-light material, with the aid of the Angels. It is only reached by man on earth, or out of earth, by his striking a tone in creation which is a totally perfect part of the Source in Creation. Those on earth closest to the Father are the ones who are doing His work out of a purity within their own soul-grace.

There are seven steps or parts to each plane in the astral world. The four lower parts of each plane retain the earth reflection, and the three upper parts of the plane express, in some degree, the light from Heaven..

The student of the "inner meaning within" is so confused over the many interpretations of the astral world that he ignores its importance to his spiritual life on earth; he overlooks the fact that he is never out of the astral world. The astral world is the first world one is trained in when the earth body is put aside. The etheric body is the body for momentum in the physical world; the body for equilibrium in the astral world; and the body for poise of the soul.

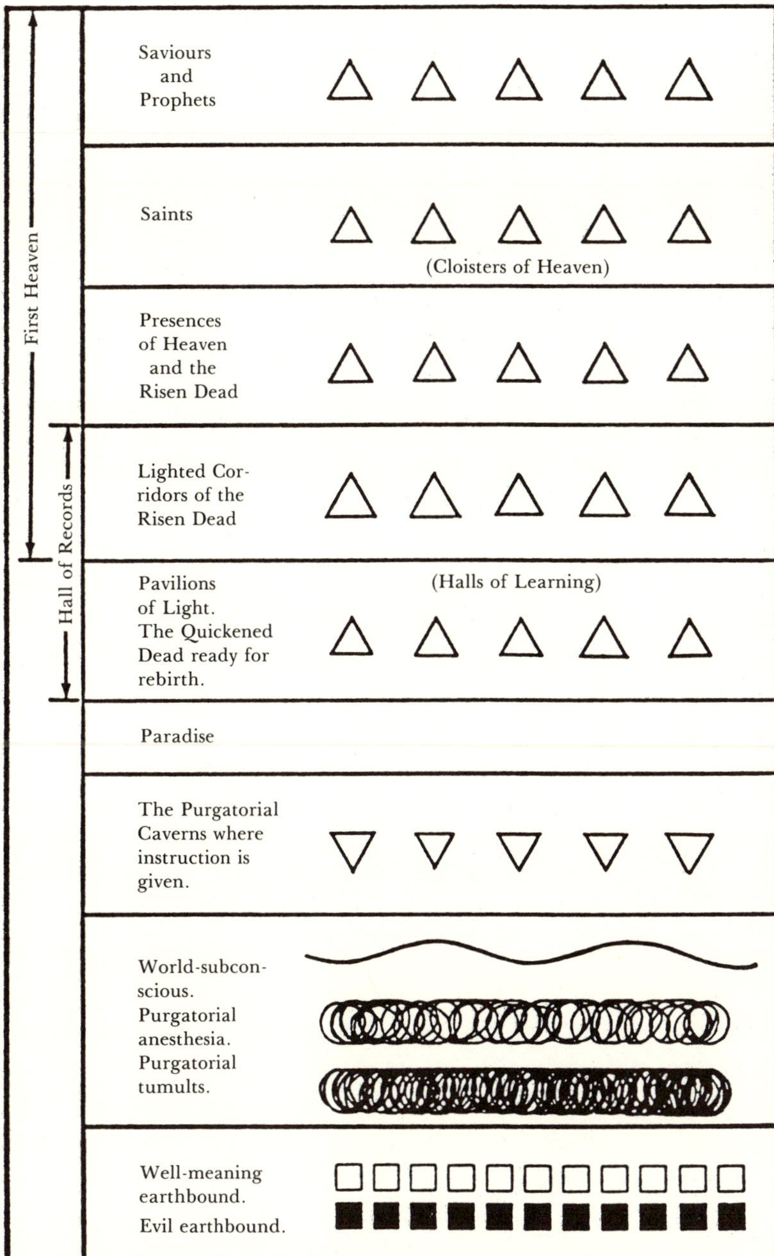

THE ASTRAL WORLD AND THE AFTERLIFE

In one's karmic turn-back, he is put in the place of those he has hurt. After death, while pain and earth-passion still cling to the pain shell, many things of a revolting and unhappy nature may be seen. This is the place where things are taught by direct example, with feeling fully shared by looking on at the suffering of others. This is a purgatorial experience and the means by which karmic detachment is reached before one can go on to higher regions to work.

This can be experienced while still on earth by dream visits or night flight. When this occurs, karma is being speeded up in both worlds—preparing one for some specializing soul-process.

Time is a mercy agent in the law of karma. Were it not for Time, men would be overcome in the overloads of force and pressures. Time is a regulating providence, giving men their chance to make their world at one with the World of God.

Deafness comes to men out of old karma. It comes from three lives in which the soul-tone has been denied, and brings into earth a temperament of stubbornness and unrelenting rigidity. In total deafness, the ear is closed off in its cavity and canal due to karma of race, family and blood. The denial of the karmic pictures seals off the hearing in the ego who refuses the responsibility within the repeated soul-tones suggesting that his karma may be worked out in action.

Those who are deaf in earth—either by birth, by injury or by chronic sickness—are fearful in life of the new and of change. Deafness is the last trial on the test of the straggler. He who fails in the softening of temperament after deafness lingers in the twilight of the astral world awaiting the birth through one who, in love, shall guide and lead him back into a way of life in which strength and courage will come.

A lukewarm view of the spiritual life is harmful. It fails

in the rhythm of activity in creation—and a biased thing is created in the thought. This becomes a partly concave formation, and is more belligerent than the strongest particle thrown by deliberation. There are astral walls surrounding people possessing such views. These walls dissolve only from inner pain. Pain is fire. When the astral walls are consumed by the fire of pain, the Presences in Light can come forth to help.

In the life after death, something of this kind happens, too, but it is not quite the same—for walls of bias and prejudice are spiral-like in the astral world; and one meets them one at a time. In the earth, the walls of bias and prejudice fall "as the walls of Jericho." Few are prepared for the exposure of their sensibilities to the downpouring of light shed on them—and an initiation occurs. At this moment the one on earth begins the work of his soul.

When men think of the astral world as an indefinable mystery, this becomes an obstruction rather than a help in their spiritual research on earth and after death.

When one is in the astral world, he cannot see the earth, but he can sense what is going on in the earth. If one receives training before death, this becomes a life-thread or line between the astral world and the physical world.

The astral world surrounds and also penetrates the earth. The lowest portions of the astral world, residing within the earth, are called the interior and anterior regions of the astral world. These lower regions of the astral world consist of a heavier ether than the ether within the higher astral planes.

In the lowest anterior regions of the astral world within the earth are undissolved prototypal shells. A prototypal shell is a discarded lower emotional body of one who, after death, has refused to be contrite for his sins. These prototypal shells are weighted down by the gravity action within the anterior level of the astral world, called purgatory. Here, men experience purgatory after death; and they also may observe it in dreams.

Fear of death binds one to the purgatorial levels in the astral world after death.

The subhuman exists in the grotesque level of the lower devachan planes of the astral world. On this level dwell also the earthbound dead, the comatose dead, and the wrathful dead who have through perversion failed to reincarnate. The wrathful dead are fixed in a shadowed purgatorial level from which they cannot escape. Having bound themselves to the fallen Lucifer influences while in the physical world, their hatreds and dark evils flow mirage-like into the minds of the living who use drugs, alcohol, or are engaged in sexual perversions.

The path of drugs is not the Virtue Path; it is not the real; it is the path of the lower astral world.

The quickened dead are in the Pavilions of Light, which are located or situated in close proximity to the First Heaven. The quickened dead behold the reflections of Heaven and intuit the Heavenly Presences nigh them. They are aware of the Angels and the Musics of Heaven, but they do not experience them in the same manner as the Risen Dead. The Risen Dead are within the light; the quickened dead behold the light.

In the Pavilions of Light, there is neither sorrowing nor grieving; the quickened dead experience a joy reflected from the higher heavenly states. Many of the mystical reports of Heaven relate to the Pavilions of Light. The golden streets, the gates of Heaven, the angelic presences, the music —all of these relate to the Pavilions of Light.

When one who lives in the physical world has spiritual penetration to a Pavilion of Light in his dreams or during sleep, he moves among the inhabitants of the Pavilion of Light. He takes upon himself the joy of this lighted place. He is aware of a state of harmony, of peace—and more, of an anticipation; for the quickened dead in the Pavilions of Light are preparing to return to the world, and their uppermost feeling is a joyful anticipation of birth.

MANTRAM TO OVERCOME THE FEAR OF THE LOWER ASTRAL WORLD

I believe in God, the Almighty,
and in Christ Jesus, His Son.

HOW TO IDENTIFY YOUR DREAM EXPERIENCE

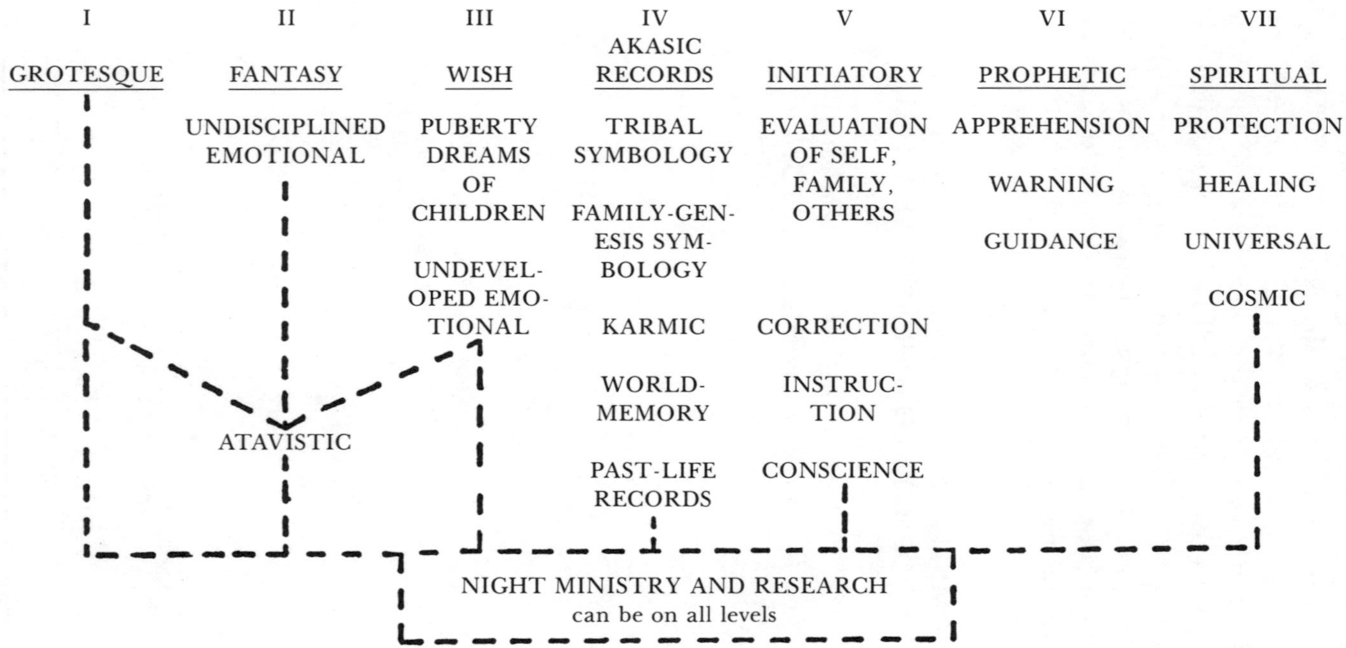

11

DREAM LEVELS

In a dream, in a vision of the night, when deep sleep falleth upon men, in slumberings upon the bed; Then he openeth the ears of men, and sealeth their instruction.

—Job 33:15,16

Initiatory Processes in the Night

Archetypal Kingdom (Spiritual)

The Christ.

Archangels. Logos Angels.

Spirit of Truth.

The Greater Archetypes or the Unmanifested Word.

Realm of Light (Spiritual)

The Lord Jesus and His Disciples. The Celestial Angels. Dream initiation in the Realm of Light is extremely rare, and is experienced only by a perfected Telepathic Disciple or a Cosmos Disciple.

Spheres of Light (Spiritual)

The Great Immortals or Masters. The Terrestrial, Seraphim and Planetary Angels. Hall of Wisdom. Dream initiation in the Spheres of Light may be experienced only by perfected Night Servers, or by those who have commanded the lesser astral planes and the 8th sphere or greater abyss trials set up by Satan.

7th Astral Plane (Sacred level)

Devas or Shining Ones.
Buddha.
Bodhisattvas.
Sacred men.
The Illuminati.

6th Astral Plane (Prophetic level)

Prophetic certainties. Communion with the Saints.
Hall of Learning.
The White Brothers.
The Risen Dead.
Cherubim Angels.

5th Astral Plane (Initiatory level)

Initiation into the abstract plane of the astral world. Contains the reflected formulas correlated to the Greater Archetypes in the Archetypal Kingdom. A threshold plane in preparation for going into the higher levels. The Labyrinths or Pavilion of Light.

4th Astral Plane (Archive Memories and Embodiment Records level)

To make alignment with the Seraphim and Cherubim Angelic Kingdom, Maha Chohan, Melchizedek, Ancient of Days and the Great Rishis, the disciple must master the three lesser astral planes. In this, he is given the power to research the Archive Memories and Akasic Records or Reincarnation Records. The beginning of the greater abyss trials.

3rd Astral Plane (Wish, Bliss and Paradise level)

To master the wish, bliss and paradise level, one must deglamorize desire; his desire must correlate to his true need seen by "Our Father which art in heaven." The

Transitory Dead are on this level. Phenomena and illusion. Predictive possibles. On the first three astral planes, one undergoes the lesser abyss trials.

2nd Astral Plane (Fantasy and Purgatory level)

To master the Fantasy and Purgatory level in dreams, there must be organization of thought, and the ability to rationalize the Plan of God.

The Unrisen Dead are on this level.

1st Astral Plane (Grotesque level)

Discarnate shells. Subtle entities. Sub-elementals. To master the Grotesque level in dreams, one must have perfect faith in the Father and His Image for man.

Until one has earned the grace to receive from the Higher Worlds, all inner experiences, such as dreams and visions, are colored by the astral planes. The lower astral planes are chimera-like, duplicating the emotions and feelings. The higher astral planes, or the First Heaven, are plateaus of inner-world governings where dwell the Lesser Saviours, the Sacred Personages, the Saints, the Elect, the smaller and greater Cherubim Angels, the Risen Dead, the Men in White Apparel, and the Higher Adepts. All good and pure persons on earth touch the First Heaven in their visions and dreams.

If one has immature emotions or traces of egotism, dream experience is distorted and discolored through psychical function stemming from the lower astral world.

All things seen in the lower astral in color are turgid, dark, nebulous. Astral color especially affects the saliva content in the mouth during dreams. An experience at night with astral color and turgid movement caused by astral dreams induces dryness in the mouth; one awakens with a bad taste in the mouth.

If one dreams that he is clearing his nose or sneezing, this is a symbol that he has stepped into inverted currents of the astral world, and thus has opened himself to dangers of attack from the dark.

In etheric anatomy experienced during dreams, the initiate learns that the eight sinus cavities are horizontal orbs and astral defenders of the body. When the etheric progenitors or envelopes supporting the sinuses are weighted down by sub-sensory energies of the astral world, the result is ineffective action in the physical sinus. To suffer continually from sinus pressure indicates that one is in some manner exposed to astral horizontal bombardments, and thus defenseless against the harsh ebb and flow of sub-sensory movement.

When one has mastered the astral sea, he becomes a protege of the First Heaven. He comes under the protection of the Cherubim Angels, the Saints, and all benevolent Holy Beings who have lived formerly in earth bodies.

When one dreams through the third-chakra psychological aspects, he goes no further than the higher astral planes or First Heaven. Superconscious dreaming takes one beyond the astral planes into the Second Heaven—and, in cases of developed adeptship, into the Third Heaven. Superconscious dreams are pictorially imaged into planes of the mind beyond the psychological.

> *The Greater Archetypes speak into superconscious dreams.*

NIGHT-FLIGHT AND THE ETHERIC LAYERS OF THE EARTH

Etherically, the earth is like a great amphitheatre. In night-flight one begins to be initiated at the outer circumference—going inward, layer by layer, into the earth, until he reaches the innermost core. As he moves inward, he becomes more and more detached from gravity. When one reaches the core of the earth, the velocity and the higher degrees of light within the core of the earth act as a springboard, and one is instantly catapulted Heavenward in his higher etheric body.

When a person reaches the First Heaven, he views, in a different vision or light, the soul dramas inside the earth and on the earth. Finally, he reaches the portal of the Inner-World dramas, and he becomes a player rather than an observer.

By mastering gravity, one becomes initiated into the freedom of his higher etheric body, as he must prepare to sustain his flight and his penetration into the Inner Kingdoms. The initiate is instructed and protected throughout these initiations by the presences of the Angels, and also by holy and worthy Presences.

When one has completed initiation into the thirteen etheric layers inside the earth, in his night-flight action he becomes fully aware that he is released from his physical body each night, and that his etheric body is the body by which he is transported and travels.

The thirteen etheric layers within the earth have separate functions. The first five etheric layers support and animate all seed, root and germinal life. The next four etheric layers influence all mineral and chemical life, such as the ores and oils of the earth.

The ninth etheric layer of the earth is under the command of the mighty Archangels. The Archangels use the ninth layer to activate their heavenly compulsions during the four cosmic seasons.

The tenth, eleventh and twelfth etheric layers within the earth consist of multiple degrees of quickened light, for they are reflectors for the Christ Light centered within the thirteenth etheric layer of the earth.

If one has any remaining sentient-atom action within his emotions—and has earned initiation grace—his sentient atoms are neutralized upon the tenth etheric layer within the earth, that he may view with accuracy and dispassion all forthcoming spiritual pageantries and dramas in the First Heaven.

In one's night-flight, as he touches the eleventh layer of quickened light within the earth, his higher emotional atoms are quickened, enabling him to be more responsive to what is felt and experienced in night-flight.

The twelfth etheric layer within the earth enables one in night-flight to be united with the Heavenly Presences who come forth to accompany him and protect him while he receives his instruction for the night. Here, in the twelfth etheric layer of the earth, the intensified light prepares him to open the door to the heavenly portals before him. On this level, his sacred and eternal atoms are united with the spiritual atoms of his mentors and instructors. Henceforth, what he sees will be colored by the thoughts and wisdom of those who assist him and instruct him. He becomes co-atom to them during the period of night-flight. His research of the night is thus sustained for a more lengthy period than he would be capable of sustaining alone. With his night instructors and his Guardian Angel, he enters the thirteenth etheric layer of the earth, and he is catapulted without harm over the third abyss or eighth sphere, that he may enter into the First Heaven and become a witness to the events of Heaven.

The thirteenth etheric layer of the earth, under the command of the Christ since the Resurrection of the Lord Jesus, is the door to Heaven. All who participate in heavenly ritual must enter through this holy, sacred, heavenly door.

One receives the power to "go in and out" when he has mastered the thirteen etheric layers of the earth through initiation. As time passes, this power is also experienced and outpictured by the higher initiates in daytime action and thought.

THE THREE ABYSSES

Hell, or the sin-body of the world, is made up of the reflected sins of millions of men who have violated the purity of their souls.

The sin-body of the earth contains three separate abysses. Every negative, willful, scornful, mocking, malicious, evil, murderous thing inflicted by man is reflected and recorded on one of the three abysses in the sin-body of the earth. The reflecting ether supporting the three abysses of hell is photographic.

All negative, erroneous actions are reflected in the first abyss. All fearful, resentful, obstinate feelings are registered on the second abyss. All lustful, covetous, avaricious, greedy, doubting thoughts are reflected and pictured in the third abyss. Only with an eye of light fixed on Christ may one move over any of these three torrential etheric reflectors.

When one is ready for spiritual service and authority, he undergoes the three abyss trials and initiations. All greater initiates and saviours of the earth have consummated these trials. The Lord Jesus experienced the satanic trials while fasting in the desert. In the latter and greater trials, one is sustained by "Our Father which art in Heaven." The Angels give succor with a very special kind of ministering.

The first abyss, a rim of excruciatingly horrible sound, is located between the eighth and ninth etheric layers within the earth. If one has blood guilt or has an amoral nature, he may travel no farther than the rim of the first abyss, where the Cherubim Angels challenge him.

In night-flight, before the initiate reaches the core of the earth, he must cross over the first abyss. This abyss between the eighth and ninth etheric layers of the earth is a floating hell made up of discarded prototypal shells. These prototypal shells are the undissolved lesser emotional bodies of men who have lived in the earth and refused to face their evils and their sins. These shells are not animated by consciousness; however, they sound discordant, harsh, repelling tones, causing unspeakable tumults. The prototypal shells also have the power to draw an unprepared initiate downward into the first abyss; thus, in night-flight, the initiate may experience devastating horrors. If one is unprotected, over-curious, or precocious, he invites such experiences.

If one is protected and insulated by his Guardian Angel, he crosses the first abyss without fear, and he enters into the fiery core of the earth under the heavenly protection of Peter, the Disciple of Jesus. Peter holds "the keys of the kingdom of heaven." (St. Matthew 16:19) His influence permeates the tenth, eleventh and twelfth etheric layers of the earth.

To move over the first pit of darkness, one must have spiritually earned angelic companionship. To move over the second abyss, one must undergo forty-nine purification emotional trials, and thus reach the First Heaven through the help of the Saints. To move over the third abyss, one encounters Satan as a dark being and comes face to face with Satan; and through the help of Holy Presences, one exorcises Satan and gains the power to enter into the Second Heaven.

The first abyss is laden with the guilt and the sins of man's physical omissions, with the sins against Nature, against the animal kingdom, against all things having form. It is also purgatory-laden with the reflected pictures of the undissolved vapor shapes of the earthbound dead. The first abyss resounds an echo of the lower purgatorial regions; this gives off an horrendous and frightening sound.

In night-flight, before the initiate may reach the First Heaven, he must cross over the second abyss, which is located on the inner rim of the thirteenth etheric layer close to the core of the earth. If one has sinned against the Holy Ghost, he must contest the sub-elemental forces dwelling deep within the second abyss. If one is without ethic, or has persecuted the Saints, or retains any atheistic or agnostic thoughts and feelings, or if he has desire for magical powers, he is drawn downward into the horrendous action occurring in the second abyss, and he is challenged by the sub-elementals and also by the cynical unrisen dead.

The second abyss of contorted movement and agitation is stained and colored by the tumultuous emotions of men on earth. All that men feel negatively is imprinted upon this abyss. Inverted feeling saturates the second abyss; these chaotic feelings and desires return to disturb the feelings of the initiate when he seeks to rise to the First Heaven. Only with pure desiring may one master the effects of this abyss. The initiate experiences these abyss recoils during dreams and during waking experiences.

The last and most terrifying abyss to cross is the satanic derision abyss of the *eighth sphere*. This must be crossed again and again before one may become an initiate under the Lord Jesus. All great servers of the light must cross over the third abyss or eighth sphere until the buoyancy of their own spiritual light gives transcendency to their spiritual bodies.

The third and last abyss is crossed after one has extended the range of his atoms through initiation beyond the fantasies and chimeras of the lower astral world, and the etheric layers of the earth. Before one may transcend the third abyss and reach the First and Second Heavens, he must undergo the satanic trials through the Luciferic or fallen angels. To pass over the third abyss, he must be free of all mental and spiritual pride. In these trials and initiations, he gains the power to exorcise evil spirits, to be a worker of miracles; and he attains manifestation and de-manifestation powers. The third abyss must be crossed before one can give the Greater Archetypal Truths, or Fourth-Dimensional Wisdom, to the world.

During the satanic derision trials, initiates are sometimes disgraced in the physical world, persecuted, mutilated, demoted, ignored. All who pass over this portion of the sin-body of the earth have reached the Second Heaven while living in the body. Such initiates are thereafter chosen rather than called.

If there be any mental or spiritual pride, one is tried again and again until he comes humbly before the Throne of Grace where reigns the King of his soul.

The inverted light of the physical Sun, in conjunction with the lower lunar rays falling into the interior of the earth, forms a shadowy body or a photographic reflector for man's sinful acts and emotions. Two of these shadowy places are etheric serpentine-like corridors inside the earth. The inverted light of the physical Sun and the lower rays of the Moon produce the first two abysses in the sin-body of the earth.

The earth's own gravity shadow forms the anterior, fearful body of the eighth sphere, or ring of outer darkness. The gravity shadow of the earth produces the third abyss or exterior portion of the sin-body of the earth.

These three shadowy places are influenced by Satan and his leagues—the souls of evil persons, the dark or fallen angels, the elemental subtle forces, the earthbound dead.

Men on earth visit and observe these shadowy places in their etheric bodies during certain dream initiations—and observe and sometimes experience them after death. Telepathy from these shadowy places falls into the thoughts of man during sleep and in the waking state. The more spiritually aware one is of the invisible worlds, the more he must encounter and master the subtle adversaries of the shadowy worlds. The only way he may master the adversary tumults is to master himself within his own thoughts, emotions and actions.

When one uses the Christ evocation or command against Satan, he may dissolve the reflected abyss pictures falling upon his thoughts and feelings. By saying, "In the Name of the Christ, go back from whence you come; and may the Father rebuke you," such pictures are automatically dissolved by the spiritual initiate. It is one of the tasks of the initiate to overcome and master the mirages outpictured from the mind of Satan. This he can do only through the mind of Christ in him.

The three abyss initiations are spoken of in the Heaven Worlds as the three greater initiations. All other initiations in night-flight pertain to one's personality, one's family, one's conscience, and to the receiving of certain instruction.

When one has fulfilled his initiations into the thirteen etheric layers of the earth, he is then ready to be initiated into the First Heaven. Should he have the grace in his soul's record to have been a gifted or talented person in former lives, he will come under the direction of the Higher Adepts who dwell within the First Heaven. The Higher Adepts will thereafter inspire him through non-intrusive telepathies to create for God. Such creations will enable the disciple-initiate to manifest a tangible, moral formula, and to be the source of illuminative and provident objectives which produce humanitarian helps and benefits for humanity.

The Higher Adepts do not as a rule name themselves or identify themselves to the disciple-initiate. Their inspiring telepathies fall into the mind of the disciple—and the disciple often feels that his own will is directing his compulsion to create. However, in the night during night-flight instruction, he is fully aware of the Higher Adepts, and he recognizes the ones who have taken upon themselves the responsibility for his instruction.

The Higher Adepts are chaste, pure, creative persons who have lived in the earth and have left their works as tangible reminders, that men may know they are more than the beasts of the field. Their presence in the disciple-initiate's life is devoid of excitation. Their labors consist mainly of keeping the lamp of inspiration and creativity ever before the vision of the creator-to-be.

The Higher Adepts have been sages, statesmen, painters, sculptors, writers, poets, composers. When one is ready to be a novitiate under a Higher Adept, he comes under the direction of the greater Cherubim Angels. These are the Angels who watch over the boundaries of the First Heaven, and prevent entry into the realms of Heaven by those who come with soiled hands or irreverent hearts. No one can cross the boundary of initiation into the First Heaven unless he has been shaped, molded, fired and purified by the Cherubim Angels.

A white lion seen in a vision or a dream symbolizes an adept; one who has command of his emotions and mastery of the astral regions and planes; also a Master. A brown lion symbolizes one who is at home in the higher astral world; one who has physical-plane manifestation powers.

To see an eagle flying indicates the power to ascend above the astral world.

A swan symbolizes mastery over the astral world. A white swan represents the levels of the Saints. A brown swan represents a person who has the power to receive certain practical helps from the Hall of Learning. A black swan denotes a black occultist or magician posing as a peace-giver.

THE EIGHTH SPHERE

> *The eighth heartbeat of man enables him to respond to the base clef tone of the earth, and thus to know fear and pain. The eighth planet in constellation is Pluto, the planet ruling the subconscious mind of man. The eighth process of thought is inverted or psychotic thinking. Jesus used His healing power to cleanse out the evil spirits in the insane men, because He had command over all dominions, all regions, all planes, all spheres.*

There exists a darkened abyss or strip of discordant sound between the third and fourth planes of the astral world. The darkest rim of this abyss is *the eighth sphere*. The abyss, as a roar of discordant sound, is located beneath the lower boundary of the First Heaven. Each night the server in Light must move over the abyss-barrier, so as to rise to the Light of the night. Until he has mastered the abyss, he is assisted by either the Angels, the Risen Dead, the Men in White Apparel, the Illuminati, or the Masters. The assistance received is determined by his evolvement.

The third abyss, or eighth sphere, is outside of the earth. The eighth sphere is a darkened region, or gravity shadow of the earth, obstructing one's entry into Heaven. Jesus called this region "outer darkness." In the lowest layers of the third abyss, or eighth sphere, dwell the withdrawn or twilight ones who are restrained from re-embodiment, and who will remain so until this eternity is completed.

The one called Judas, who died contrite, is in command of the eighth sphere. Here, he seeks to reinstate those who have fallen and descended into the darkened part of their souls' light. He works to enable those who have detached themselves from the record of their souls to repent and to be retrieved. After death, all utterly depraved forms of life dwell within the eighth sphere.

Until this day that which is known as the eighth sphere has been feared by all knowing the workings within the inner planes. The eighth note of the octave, the eighth planet in the constellation, the eighth heartbeat of man, his eighth thought process, and the eighth sphere within the physical earth corresponding to the astral world—these have remained a mystery sealed away from man. All in this earth having been created in the 7, that of the 8 has withheld its secret. Average man has not in his vehicles the capacity to maintain his poise, his rational will and mind, in the study of the 8.

All advanced initiates—who from life to life have worked with the Higher Worlds—have known the secret of the eighth sphere, the eighth tone, the eighth power; they have known the Wisdom which is that Light within Light. He who stands on the threshold of stepping away from the wheel of karma has entered beyond the eighth sphere; has mastered it; has overcome that which is elemental in the earth; has come to understand traveling in the night with freedom; has risen above the darkened tones in the lower astral regions; has in past bodies known the laws of levitation or lifting of his vehicles; has known the power of luminosity or disappearance; and has known the lifting of matter. He who has commanded the eighth sphere and mastered it has never commanded it for selfish will or for self.

All persons who tamper with the unknown forces of the inner planes—through selfish will—touch the region of the dark. On touching the eighth sphere, there is set loose a pandemonium, in which drastic conditions come forth from the darker regions.

The nebula within the nebula has collected out of the ages—from one chaos to another—the discarded shells of men, discarded thought patterns, discarded evil deeds, discarded forms and frames of men. These gather in the eighth sphere awaiting the day of the recalling into Light. These forces touch on men who misuse psychic force through selfish will. These are the denizens of the dark. These are the fearful forms which come forth and dare men to cross the threshold into the Higher Spheres.

It is with clean hands and a pure heart that men rise into the higher regions of Light, reflect on the spheres of the planets, and see their work in the coming Archetypes for men. Each planet—a part of the constellation around the earth, living in the virgin light of the Sun—carries its secret unto the earth. All forms to be created are first placed as in carbon print within this reflecting light around the earth.

The sphere of Neptune, the sphere of Uranus, the sphere of Saturn, the sphere of Mars, the sphere of Venus, the sphere of Jupiter, the sphere of Pluto—all carry their secrets of that which is yet to come. Those carrying with them the mark of Pluto in their foreheads shall penetrate above the eighth sphere, gather knowledge within the Constellations, and garner treasure of that which awaits to ripen to men in their work in form on earth.

The Moon and the Emotional Body

> *The Moon's reflected Sphere is activated in the lower and the higher astral world. When a person is initiated in the Moon's reflected Sphere, the astral world is clarified to him, and he masters the currents of the astral world.*
>
> *In the initiations within the Moon's reflected Sphere, one learns of his emotions.*

The *astral body* is the emotional body.

The *astral core* is an ovoid-shaped envelope of astral fire kept in balance by nadi points or sound vortices located along the spinal system. The nadis act as absorption points receiving the sound and vibratory currents from the planets. The astral core represents the emotional body.

The planetary fire playing upon the spinal system is distributed and generated by the assistance of the glands and the nadi points. Thus, the nervous system is protected in the etheric and physical bodies from being the recipient of direct vibrational force playing upon the spinal canal. When the astral core is unprotected, the glandular system goes out of balance, and the emphasis is an overcharge of psychic energy in the etheric body and in the lesser mind. One protects himself from the unruly charge of the astral-core fire through mantrams, contemplation, meditation, prayer and selfless service.

A lunar psychic is one who is engrossed psychically with his emotions and sees all through emotions and feeling. He is dependent upon the astral lunar reflective light for his psychic powers. The lower lunar psychic is unable to interpret what he sees. The higher lunar psychic sees in part.

The lunar and solar psychics are initiated through the planetary energies playing upon the astral fire in the astral core along their spinal systems.

The glands of the body are planetary receptive generators. It is the work of the astral core to send the charged planetary energies into the etheric body.

A higher lunar psychic is an astral psychic having access to the reflective light of the lower astral world and to the higher aspects of the First Heaven. Lunar psychics, both high and low, use their power of ESP to see all things clairvoyantly in *reflected* light. They are profound night dreamers of the weird, the fantastic, and the magnified. Lunar psychics use the power of psychometry, of character analysis, of physical body diagnosis. The lunar psychic is responsive to hypnotism by persons in the physical as well as in the interior subtle worlds. Lunar psychics use their powers to unite with the Moon changes working through families, the animal kingdom and Nature.

Man, through his astral nature, is exposed to sound, to movement, to feeling. He learns of this through myth-dramas functioning within the astral planes. Through moods of longing, yearning and pure desiring, one unites with the higher astral planes and comes in contact with the emotional body of the world, or the Buddha-chain supporting the emotional body of the earth.

One who thinks wholly with the lower astral nature is out of the range of higher self-awareness. One who thinks wholly with higher astral thoughts is prescient when the higher ego or self is in command. To think in total lower thinking, or astrally, one becomes a subjective psychic receiver. To understand and to use the resources of the higher psychic nature one must objectify and command the glandular star-center higher velocities located in the spinal nervous system. These centers are stimulated by the solar and lunar fires working with the varying changes of planetary action.

If one is oversubjective, impressionable or suggestible—that is, plus-astral—he is not aware of himself as a whole person. He is content to emotionally respond to life; he is a fated vessel rather than an initiate.

The astral psychic nature in its lowest aspects produces the exploitative psychic, the witch, the mediumistic seer living vicariously through the magnified ectoplasms and telepathic compulsions of the unrisen dead. The harsh side of the Moon or lunar tides works with the lower aspect of the psychic nature, using the chimera reflective lunar light rather than the higher direct *solar light* supporting the planetary or starry points in each gland. The glands of the lower psychic express the lunar astral nature; such psychics work with the gravity tides and the lower electronic frequencies of the earth and the lower vibrations of the planets intermingling with the magnetic belt surrounding the earth. The solar-tides initiate works with the higher astral refined, supercharged wave lengths. The sun or solar-light initiate is under the command of the Christ Spirit. The lunar astral-light psychic is under the command of the Race Lord, Jehovah, and the Lunar and Terrestrial Angels.

Thus, a person of predominantly lunar nature is astrally psychic and is gravity bound to the acts of phenomena produced by the astral world. The contrasting tumults in the psychical lunar temperament, sometimes ecstatic and sometimes melancholy, are best analyzed by an understanding of the astral chimera variables, and by a comprehension of the intricate system of karma.

Solar initiates are spiritual psychics, having reached the fourth aspect of transcendental action. They use the higher astral psychic stuff (akasia) of the First Heaven and Second Heaven to rise beyond karma and to heal emotions and thoughts. They are fully aware of all of the psychical currents in their own thoughts and of the nuances supporting the higher psychic awareness.

In the astral world are moving, transparent, negative thought forms which are dissolved and rebuilt by men with positive thoughts. In this way the lower thought world is organized.

The astral world affects men in the New Moon through the pictures playing on men's minds in the Earth. Into their thoughts images are gathered which have been cast forth by the New Moon's rays. These gather as smoke and spiral downward onto men's thoughts. The emotions of men rise. Those who are less spiritually evolved become exhausted with the thoughts of negation approaching the Earth; but he who is in the Master's Light is raised to the vibration of purity and peace, and withdraws himself from man's affairs and stands only in the Master's Light, living in the high of his higher mind.

In the New of the Moon, one is more likely to experience a Saturn initiation. Through the Saturn initiation, the Luminosity Atoms around the Sacred Atom of the heart are stirred. When this occurs, the power to appear and disappear etherically is received. One learns of the powers of mental levitation, or of the telepathic sending of his thoughts. He gains the power to observe the inner workings of the earth, and learns of the constitution of the earth. He penetrates the little-known secrets hidden within the mountains and the streams. The Saturn initiate is free to move, etherically, in and out of the layers of the earth. He overcomes the gravity pull of the earth and is at home in the earth's inner atmosphere. Before anyone can research or learn of the upper stratosphere of Heaven, he must first undergo the Saturn initiation.

On the tenth day of the increasing Moon, there is a unique function of gravity, which gives to man certain levitational powers of the etheric body, and especially of the emotions and mind. This raising power on the tenth day of the increase of the Moon enables one to rise beyond obstacles in the emotions, and move beyond his fears and timidities. It is also a day in which re-evaluation in certain portions of the mind endows one with interdimensional aspects in thinking and in thought.

To fast on the tenth day after the New Moon is to work with Time and Space; to experience this brings an awareness of enlargement and freedom. The Divine Omnipresence on this day says to the initiate: "*Behold, I am the Omnipresence. I am come forth, that ye may live in abundance, and break free from enclosures of habits. I am Free Spirit. When thou art free, I am the Living God in thee.*"

The physical body and the emotional body are designed

to desire always more. This is an especial function of the greed related to the third chakra. In the tenth day after the increase of the Moon fast, one rearranges the desire patterns.

In the tenth day after the New Moon, one enters into the spontaneity realm, which is not habit-binding. One opens the Coinciding Principle and moves out of the labyrinths of emotional bondage to the senses.

From the time of the tenth day after the New Moon, gravity-freedom is experienced into the Full Moon. The Full Moon experience is a Yin and Yang experience, acting as a will power point of achievement for the one who regularly practices the fasting on the tenth day after the New Moon. Thus, he who observes this tenth day fasting has an experience in the following Full Moon that others not so observant do not experience.

The tenth day after the New Moon is the crossing point between Yin and Yang. Yang, as will, becomes predominant in the spiraling Sushumna electrical fire within the spine of the initiate; he becomes a power Aaron's rod with full budding for God.

The two hemispheres of the brain are united in the tenth day after the New Moon. The bud, the flower of the greater lotuses or the upper chakras, seeks to levitate into full bloom. One walks physically lighter with less gravity pulling the body down to old age, chronic sickness and impossible situations in karma. He can be said to be Mercury-heeled or fleet of foot in all he seeks to do in the physical world.

There is a very special *Sun* work on the tenth day after the New Moon, which enables the mental planets—Mercury and Saturn—to come closer to the initiate. These planets are the two mental planets affecting man's solar nature and his mind-building propensities. Mercury and Saturn are especially free as energy powerhouses in the mind of man on this day of fasting.

During five days after the Full Moon, the subtle and inverted forces of the lower astral regions play upon the lower emotions of men.

The eleven days after the Moon's darkness touches the astral world; man's emotions turn inward and he places his questions on the Moon-tide on which the light of the New Moon brings the answers.

The Venerable One introduces man to the Devachanic Kingdom, to the First Heaven.

12

TELEPATHY AND HEAVENLY BEINGS

Believe on the greatness of thy soul so that thou might unite thyself with the greatness in the souls of immortal men.

The Nervous System

Within the nervous system there lieth the greatest system of telepathy man may ever conceive. Naught of things scientific or mechanical may ever envision that which is truly to become as man's nervous system in the coming days. The nerves, which relate man to the Spiritual Worlds through the Kundalini or spinal structure, give to man the future hint of that prepared for him. His nervous system will become a great dispenser of mind, of spreading Light; a luminous Light on which the soul-light of another shall blend, and men shall know not or see not of that which they are as of today.

Men work daily in this building, perfecting and creating of their bodies. Through the use of consciousness, they constantly work with Light through the nerves, the brain, the blood, and that within the heart's system unto the new and great mind as Illumination.

The nervous system, the glandular system, and the organs of the physical body, inclusive of the heart and the brain, are subjectively responsive to the psychical aspects of the soul; the organs seek to respond positively to the spiritual aspects of the soul. The subjective powers of the soul are spiritual. It is the work of every living conscious being in the earth to unite the subjective facets of the soul with the objective facets of the mind.

Galaxy forces and energies formerly sealed away from man are now saturating the axis core and the magnetic points in the earth, and the astral core of man, playing upon his spirit, will and nervous system. With the coming of the Christ-Spirit elevation in the earth, man began a period of initiation assisted by planetary forces and energies. The display of unique expression in consciousness in this age is only the beginning of a realistic rather than mystic approach to the life of the mind and eventually to the life of the soul.

The higher aspects of the Divine Mother or Feminine Principle maintain the balance between the involuntary nervous and muscular systems of the physical body, keeping alive the vital psychic powers functioning within the psyche or soul attributes of man. All psychical energy, force and power generate vibration, energy, electricity. In higher conscious thought this reacts as light. In the lesser mind this generates a sub-electricity acting as a mental heat. When thought is wed solely with the dot-and-dash action of the voluntary nervous and muscular systems, this produces the forceful personal will.

The Divine Mother or Feminine-Principle influence upon the involuntary nervous system persists in life and in death. Divine-Mother life force within man is preservative as well as disciplinary. When the Divine-Mother action of the involuntary nervous system is channeled into the outer consciousness as selfless love, health of mind, body and soul is assured.

To materialistically permit the senses full reign, places pressures upon the involuntary nervous and muscular systems; the Divine-Mother action then becomes a disciplining and sometimes retributive action. Spiritually surrendering to the Divine Mother's inductive suggestion and direction of the higher unconscious produces a supra-sense perception or divine intuition. By this, one perceives the heavenly realities through the single or divine eye of spirit.

Above the base of the skull at the conjoining points of the involuntary nervous system, one's thoughts during meditation flow upward into the unified or blended cosmic matrix existing between the Father Principle and the Mother Principle. Here one experiences the Divine Marriage, receiving the androgynous powers in a bliss union with God.

Miraculous healing, initiatory powers, spiritual gifts—all result from communion with this holy union of the subliminal power of the Cosmic or Divine-Mother Principle and the imaging restoring will-life of the Father Principle. When one meditates, he must move beyond the lower Quelle psychic energy excitation into the exalted bliss of oneness with God. He does so by being initiated into the marriage of the Divine-Mother power and the Heavenly-Father power within the crucible of his higher mind.

> *The will and the nerves are twin systems through which the soul may raise man to a higher degree of consciousness.*

Spiritual Telepathy

Spiritual telepathy is the art of communication and communion between physical and spiritual minds on similar light-waves of life. In spiritual communion through telepathy, there can be no communication unless the communicating minds have similar or equal degrees of light. One is totally sealed away from spiritual telepathy if he short-circuits his degree of light through doubt of God as the uniting One in Will, Life, Light and Love.

In spiritual telepathy, the scale of thought-velocities

between souls may vary, but the degree of light must be the same. A Greater Being has the power to slow down the velocity and intensity of his light. Thus, a Greater Being may communicate with one having a lesser degree of illumination.

A person raises his degree of light by having an unwavering faith in God. In this manner, he can receive telepathic thoughts of light from Beings having greater degrees of light than he. However, when one has scepticism, doubt, hate, anger, fear, hostility or arrogance, he opens his mind in dreams and in waking to a horde of negative telepathic depleting, demoting and inverted ideas.

In telepathic communion through faith in God, one receives uninterrupted inspiration and guidance. In seeking to receive psychical communication telepathies by placing one's faith in mechanical objects, such as Ouija boards, crystal balls, Tarot cards, or hallucinatory drugs, one opens himself to darkness, mental derangement, psychic abnormalities.

The Life-of-God telepathies are transposed through life-restoring, life-sustaining, life-providing and life-knowing.

The Light-of-God telepathies are transposed through the Christ Mind into omniscient minds and into the higher mind and soul-light of the ego. Through the Light-of-God telepathies one receives transcendental powers.

The Love-of-God telepathies are transposed through the Lord Jesus and the Holy Ghost—reassuring, comforting, forgiving, healing; these are experienced by man through faith in God. Through the Love-of-God telepathies one is transformed.

SPIRITUAL LEVELS OF TELEPATHY

GOD—Eternal Spirit; the Will-of-God telepathies are transposed through Heavenly Beings as vibrationless powers and stillness, bringing peace and making it possible to *know* God.

CHRIST SPIRIT—Works with the Archetypal and greater-ideas telepathies; vibrational revelation.

FATHER—Works with the imaging, making and creation telepathy.

HIERARCHS—Work with the temperament and prototypal willing-forming telepathy, using the planetary, cosmic, zodiacal constellation energies accompanying our earth system.

JESUS—Works with Saviour, love, healing, forgiving, manifestation and ethic telepathies; commanding eternal cosmos forces, cosmic energies, astral currents through His etheric body as an instrument of mediation.

DISCIPLES OF JESUS—Work with the prototypal-ethic telepathies; apostle gifts, powers, prototypal ethics and soul-record grace.

THE FOUR ARCHANGELS (Raphael, Gabriel, Michael, Uriel)—Telepathically release the Archetypal ideas during the four seasons; move men out of karmic fixations. The three remaining Archangels—Raguel, Zerachiel, Jeremiel—release telepathies to increase the spiritual power of the Holy Ghost into the souls of men so that men may forget not their God.

CELESTIAL ANGELS—Work through initiates and adepts to reveal the inner wisdom-truths of God.

HOLY GHOST (Command of the Archetones)—Works with the Archetone or the sound stream, audible sound or vibration, the WORD; works with charismatic telepathies and power, signs and spiritual wonders. The Holy Ghost may be heard in dreams and meditation in supradimensional echoes as the Voice of God.

SERAPHIM ANGELS (6-winged)—Protectors of the great Archetypal ideas; in dreams and meditation one encounters the Seraphim Angels so that he may receive initiation into the higher mind. The Seraphim Angels work with the six soul-portals of the mind, opening the higher rational spirit in thinking.

PLANETARY LOGOS ANGELS—Work with higher planetary telepathic tones and Spheres-of-Light telepathies. In dreams one is initiated into the reflected planetary Spheres of Light, uniting with universal and cosmic creation. Each planet contains a variation of tones that an initiate must master. From life to life the initiate incorporates the tones of the planets through the help of the Planetary Logos Angels. The mastery of the tones of the planets gives the power of synchronization.

SUN ARCHANGELS—Work with Mary; the adept's protectors; work in the night and during meditation and in the night during dreams to free the solar fire in the right hemisphere of the brain. The chief Solar Angel is called Metatron.

MARY—Miracle telepathy; levitation powers; transmutation powers; visitation power through command of the four-body eternal atoms; the power to appear in crisis to petitioners or supplicants. To see Mary is to touch the Second Heaven and to commune with the Most High Saints who are under Her charge and the charge of Her Son. Her symbol and Her signs are more often contacted at high noon each day. To dream of Mary is to be cleansed, purified of hostilities. To see Her in a vision is to be blessed with very special sanctification powers.

24 ELDERS—Patriarch telepathy; authority for the worthy; the dual Hierarchs or Elohim working with the chromosomes of man.

ANCIENT OF DAYS—Timing telepathy; advanced initiatory trials and happenings; the benign Father Principle expressed by an omniscient progenitor of the Heavenly Father. He is seen in dreams when one is to learn something of timing and its command, and thus to become timeless.

JEHOVAH—Tribal and racial telepathies in dreams and waking; Family-Atom cleansings. In meditation, Jehovah, working over the lower aspect of the divine thalamus, sends the guilt pictures of the world onto the mind's screen of the initiate, that the initiate may covenant to give soul liberation to mankind.

LUNAR ANGELS—Propagation telepathies; recalling of lesser-etheric-body forces. In dreams and meditation, a lunar psychic using the lower psychic forces receives fearful apprehension, visions and predictions from the Sharp Angels. The higher lunar psychic receives from the Lunar Angels foretelling of true things, realistically oriented to man's vital need and comfort.

MOST HIGH SAINTS—Miraculous telepathy and works; in meditation assures the initiate of comfort, peace and protection. In dreams the Most High Saints or Bodhisattvas *take* the initiate to scenes of the record of martyrdom; most specifically the martyrdom of Jesus is revealed by the Most High Saints.

CHERUBIM ANGELS—Building and creativity telepathy; telepathic stimulation of humor; telepathy through music; healing the depressed. To be aware of the Cherubim Angels in meditation and dreams is to know that one will have supernatural helps in his development on the physical, mental and spiritual planes. The Cherubim support all servers for God by sustaining them with flawless skills.

LESSER SAINTS—Healing telepathies; healing the persecuted; may be seen in meditation as invisible gift-bringing friends from the invisible planes. Their mercy helps in dreams give assurance that one is blessed, led and comforted. Every person has a Saint who works with him in certain times of meditation and dreams. The Saint wears a duplicate face to his own, but is not himself. In time one learns by divine intuition what his Saint is saying to him.

DIVINE MOTHER—Telepathies are received as guidance to love more, to be giving, tender, to enlarge the Love Principle through the Feminine Principle. Her symbols seen telepathically are the white lotus, the white rose. Through telepathy, She guides a woman initiate to use sacredly the feminine functions of her body, such as motherhood, giving birth, the sexual act, household tasks, the handling of food, the training of children.

MELCHISEDEC—Manifestation and de-manifestation telepathies. He initiates through fire.

MAHA CHOHAN—Heavenly-governings telepathy. He initiates through organizing the Law Principle.

JUDGMENT ANGELS—Guilt and conscience telepathies. These initiate through the conscience.

LITURGICAL ANGELS—Activate worship telepathies. They work with Holy Ghost vibrations.

LITANY ANGELS—Ritual telepathies. They work with rhythm.

WHITE BROTHERS (Men in White Apparel)—Protectors of the Risen-Dead telepathies. They work with the higher emotional body.

RISEN DEAD—Protective telepathies. Exorcisms. (Note: In spiritual experience it is necessary to understand and interpret the telepathies from the *unrisen dead* in order to gain exorcism powers. Unrisen-dead telepathies relate to misdirected obstructions, karmic telepathies and dark-magic telepathies.)

TERRESTRIAL ANGELS—Wisdom telepathies regarding grains, plants and all reproducing things in man's and the animals' domain.

ILLUMINATI OR HIGHER ADEPTS—Creative telepathies. They are mentor telepathic directors over all artists.

GREAT IMMORTALS OR THE MASTERS—Instruction telepathies. They work to keep the disciple on the Path; to bring unmanifested ideas.

MAN'S PERSONAL ANGELS—Man's companions since the beginning of this eternity system.

Luminosity Angel—care of body and time-of-death telepathy.

Angel of Pure Desiring—emotional response telepathy.

Niscience Angel—pure thought and ethic-of-thought telepathy.

Recording Angel—akasic-record telepathy.

Guardian Angel—apprehension and protection telepathies.

HIGHER SELF—The eternal-reality telepathies.

SOUL—Akasic grace-record telepathy and vibratory-hum telepathy.

HIGHER EGO (Centered Self)—Past-life accomplishment telepathies.

INVISIBLE HELPERS—Healing, ethic, reassuring and prompting telepathies in sleeping or waking.

INTERCESSORY BEINGS AND THEIR IDENTIFYING SYMBOLS

Through depth initiation in dreams and meditation, one touches the symbolic storehouse of Heaven. In time, he will become familiar with certain intercessory symbols. Such symbols identify one's degree of evolvement and inform him of the source of his intercessory helps.

ARCHANGEL RAPHAEL—A white horse; a white rose; the rising sun; a sapphire.

ARCHANGEL URIEL—The aurora borealis; an emerald; a tree upside down (Archetype of Humanity Generation).

ARCHANGEL MICHAEL—A golden flaming sword; armor and breastplate; scales; a diamond.

ARCHANGEL GABRIEL—A horn; a golden horn; a red rose; a ruby; an infant child holding the hand of a man; the Madonna.

BODHISATTVA—A lotus; a cave; a diamond rod; a white serpent; a bamboo cane.

CELESTIAL ANGELS—A golden harp; a diamond.

SERAPHIM ANGELS—Six wings on the Angel; many eyes making up a medallion; a golden peacock's feather; crystal. Six-winged Seraphim symbologies have to do with ideas; the Lords of Mind or Archetypal-ideas research; an adept's vision.

CHERUBIM ANGELS—A cube; a bar of music; a musical sound heard in the inner ear; a freshet poem; all great composers receive their musics through the help of the Cherubim Angels. All great cathedrals are built through the help of the Cherubim Angels. The Cherubim Angels are the masterbuilders of Heaven; protection; manifestors for building the unmanifested; sounders of the soundless into musics. The Cherubim Angels open the wisdoms in the use of akasia.

RECORDING ANGELS—A book; a scroll; a golden chain; scales; a lesson in timing.

GUARDIAN ANGELS—A perpetual flame; a hearth; a pure white hand; a lesson in the mystery of space.

TERRESTRIAL ANGELS—A vision of the sphere of the earth; an upright tree; a human heart; a knowledge of gravity; a bird.

JESUS—A shepherd's staff; a seamless garment; a fish; a loaf of bread; a table with white cloth; a white dove; a dove in violet light; a cross; a crown of thorns; a small boat with white sails; an open tomb; a white pearl; a door; a white seamless robe; a healing hand; a sandal; a Christ child; a ring of gold.

MARY—A white lily; a mother and child; a donkey with mother and child; three women and a cross; an intercessor for a miracle.

SANAT KUMARA OR ANCIENT OF DAYS—The all-seeing eye; a golden chair; a diamond staff; a rainbow; a throne.

MAHA CHOHAN—A domed room with tiers; a screen; a red-carpeted central stairway; a tube ascending upward.

RISEN DEAD—A feeling of pure love and a remembered scene or face; a vision to warn, to protect.

THE SAINTS—A rainbow; seven golden chalices filled with anointing oil; seven golden steps; a golden slipper, blue shoe or sandal.

ILLUMINATI—A golden quill pen; a library of books; a scroll; a physician's vial; an artist's paintbrush; a compass.

THE GREAT IMMORTALS OR INWARD MASTERS

Master M—Light Stream #1. A book; a transparent crystal-like staff; a golden wedding ring; an hourglass; an archetypal tree; a diamond; a unicorn; a grain of rice; a golden bowknot, if one has earned protective telepathy; a silver bowknot, if one is under discipline; an owl. If the owl is looking east, the disciple is remembering Eastern lives. If the owl is looking west, the disciple has opened the Christ Light within his mind. If the owl is looking south, the disciple is yet bound by ancestry. If the owl is looking north, the disciple is to be sent to a new polarity, requiring him to pioneer in a new spiritual theme.

The Venerable One—Light Stream #2. A white staff; a pink rose; a white dove flying in rosy light; a ruby; an acorn; a grain of corn.

Master Serapis—Light Stream #3. Solomon's Seal; three interlinked loops; an amber stone; an amber staff; a dove flying in orange or amber light; the Star of David; a stalk of millet grain.

Master K.H.—Light Stream #4. A jewelled throne-chair; Hippocrates' symbol or a serpent on a cross;

a red rose; a surgeon's bowl and white towel; an emerald; a dove flying in etheric, Neptunian green light; a grain of rye; indigo blue light.

Master Hilarion—Light Stream #5. A white butterfly; symbol of Uranus; sapphire; a dove flying in sapphire blue light; a grain of oats.

Master R—Light Stream #7. A white rose; an African violet; scales; pyramid; obelisk; an eagle; phoenix bird; an amethyst stone; a dove flying in amethyst light; a key; a grain of wheat; a bunch of grapes.

A TEACHER OR SACRED GURU—A pelican; a snowtop mountain; a string of pearls or jewels. A teacher clothed in a white garment surrounded by diamond-like light is an etheric guru of the highest order. A teacher with the lower part of his garment colored from the rosy earth ether, the upper part of the garment diamond-like, is a teacher still living within the reach of gravity, as in the First Heaven and Second Heaven; such teachers come only to advanced initiates.

ADDITIONAL INTERCESSORY SYMBOLS

To shake the Master's hand in a dream is to be accepted as a probationer; to make a covenant to accept one's karmic cleansing and thereby render a service to aid in the soul liberation of others.

When the Master kisses one on the cheek, the one being kissed has become an accepted disciple or initiate. The initiate should be prepared thereafter to render the kiss of communication and salvation to all others.

Vision of Great Being at top of staircase—Master K.H. leads one up to healing work; to be trained in night healing; to blend with Hippocrates' ethic in healing; to learn etheric healing techniques. A surgeon's bowl and towel; a red ruby; a red rose; a surgeon's scalpel; geometrical designs; the initiate is being trained in the techniques of etheric anatomy and solar-light healing.

To turn right and to walk up a spiral stairway means that one is preparing to meet Master K.H. and to be instructed into the 4th Light Stream of healing the impossible; also is preparing to move into higher planes for self-research and akasic soul-record research.

Walking downstairs—a memory of returning to the body after sleep.

To see a blue light while meditating means one has made contact with one of his Angels close by. He also has seen a portion of his spiritual-atom portal open in the throat or between the brows. This may mean that one is making a contact with a pure Presence of the Light, such as an Angel, an invisible helper, a higher guru, one of the Masters, or a beloved Risen Dead. One should always give thanks for this vision so as to extend the time of the blessing from the Presence.

Rainbow—Cherubim; a saint's protection; also Ancient of Days; the power to rise to the First Heaven and receive help from the saints.

Teacher or a Being with feet on floor—means the one seen is living in the earth. If his feet are above the floor, he is in the etheric or omniscient state. If he is seated in a yoga posture, this is the sign of Master-omnipresent powers or being permanently conjoined with his protege apprentice in the earth.

Feet on earth—earth worker and earth teacher; also invisible helper and night server with visitation powers.

Feet above ground—teacher with ascension powers in the spiritual worlds. Such teachers give to the earth initiate levitational powers, that he may rise into greater ranges of light, and lift those in the world from their sicknesses, griefs, sorrows and awkward ways.

Breath in a vision or dream—the Holy Ghost; uniting with prana or cosmic life-force; the need to unite with the pulsation point of the soul's medallion so as to be at one with the vibrationless power of God.

Jesus seen in a vision giving off a beautiful indigo blue on the inner rim of His head nimbus means one has made contact with Christ Consciousness.

The Eternal Spirit fire or mind's light of Jesus is always seen in a cerulean blue light of cool peace-giving, celestial fire.

Intercessory symbols may be received during dreams or daytime contemplation and meditation. One should familiarize himself with the intercessory symbols so that he may know the source and level of his spiritual training and instruction.

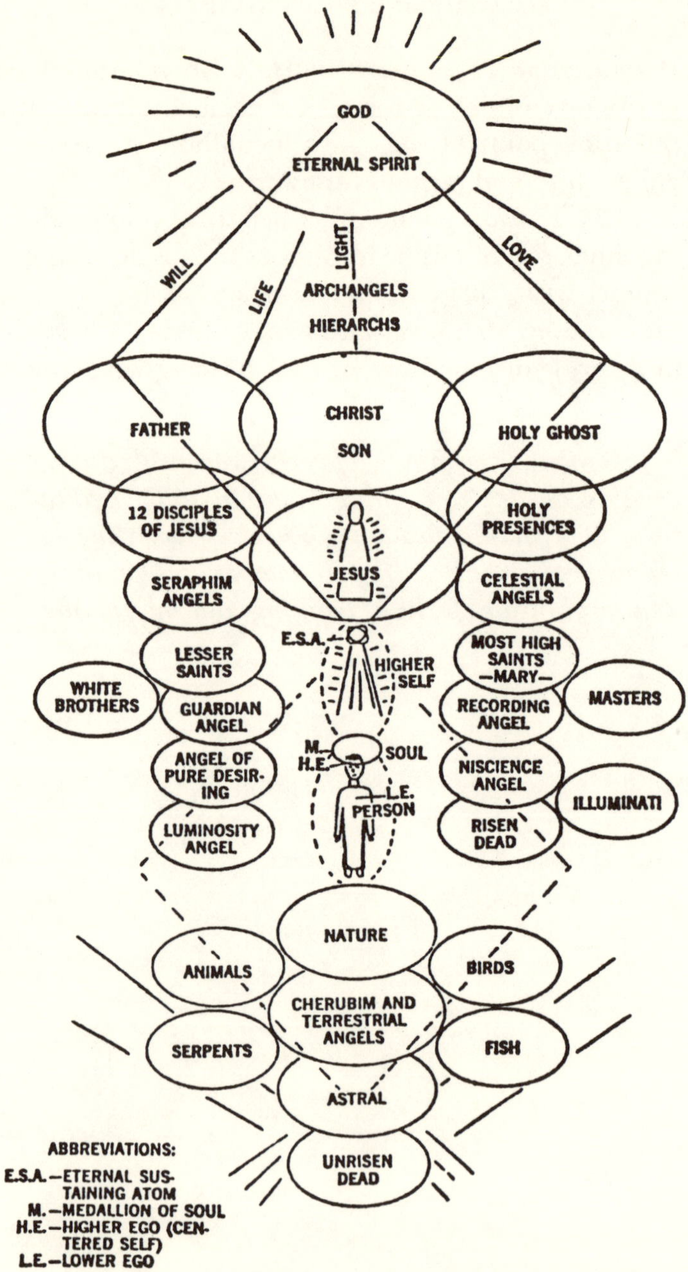

MEDIATION STATIONS

13

THE FIRST HEAVEN

In my Father's house are many mansions; if it were not so, I would have told you. I go and prepare a place for you. And if I go and prepare a place for you, I will come again, and receive you unto myself; that where I am, there ye may also be.

—St. John 14:2,3

Heaven

Men reach Heaven not by difficult climbings, but men reach Heaven by the simplicity of the giving of the self in love to the One. When one placeth self into the Light, he magnifieth the soul's identity and blendeth with the One.

Men do not by noise or by the making of magical tones reach Heaven, or view Heaven, or receive of Heaven's Light. But men, by love, create the tone by which they travel—the tone by which the Angels make possible the passage between Earth and Heaven.

Billions of years have set this planet spinning, and thy soul hath accompanied its coming into birth. Thy soul-age is eternal. All that is in thee reflecteth the Divine Universe.

Jesus as the Mediator for Omnipresence teaches us that there are actual places, vicinities and locations in Heaven. Each eternity system has a Heaven. Some eternity systems are living within full knowledge of heavenly energies and vibrations. Some eternity systems are totally unaware of Heaven and of Omnipresence.

All who are progeny souls for Earth systems must unite with their own portion of Heaven in their own eternity system.

Our Father in Heaven is the Hierarch of our eternity system. Through Jesus as His co-atom Son in this eternity system, the Angelic Kingdom is open to the soul of man. The Angelic Kingdom is the Third Kingdom through which one masters Duality.

Heaven is the polarizing constant for Earth-life density. All Archetypal processes in Heaven are ordained Archetypal processes or Ultimates.

In the Blueprints of Heaven may be found the Will of Absolute God for any particular eternity system.

There are degrees of Heaven inherent in man. If he has within himself good, he is responsive to the plateau of Heaven where dwell the Cherubim Angels; this plateau will enable him to achieve integrity, honor, charity. If he has within himself both peace and good, he will be responsive to the plateau of Heaven where dwell the Saints; this is the plateau of equanimity and serenity. If he has the grace to have obtained a portion of wisdom, he will respond to another plateau of Heaven—the plateau of instruction where dwell the Seraphim and the Great Immortals. As he increases in his wisdom and obeys the instruction of Heaven, he shall be given greater understanding of the Kingdom of God.

Since the Kingdom of God is within us, we must be there, aware of the Omnipresence of God as being also there.

THE HIGHER ASTRAL WORLD

> *Wherefore seeing we also are compassed about with so great a cloud of witnesses, let us lay aside every weight, and the sin which doth so easily beset us, and let us run with patience the race that is set before us, looking unto Jesus the author and finisher of our faith; who for the joy that was set before him, endured the cross, despising the shame, and is set down at the right hand of the throne of God.*
>
> —*Hebrews 12:1,2*

We are surrounded by "a cloud of witnesses." These witnesses are the Holy Presences of Heaven. Mediation with the Holy Presences of Heaven enables one to build a mighty, expanding circumference of love and to work mediatively for the welfare of individuals.

The power of Mediation, the power of Mediative Telepathy, the power of Mediative Healing, the power of Mediative help and source of strength in time of trouble—these are ever present when the disciple understands Mediation.

One cannot advance or increase alone; he cannot ascend alone. He has constant helps, continuing helps from the many Presences of Heaven and from those who have Heaven in their hearts.

All living souls are united in Mediation. Mediation is a mighty Honeycomb, a Medallion, a Filigree of Light—uniting, knitting, linking, coalescing, mingling, blending, correlating, corresponding. The communion in Mediation enables one to become at one with Him. The majesty of Mediation resolves the mystery and brings mastery to him who would seek the Source with ethic. Dedication is the absolute faith in the Presences of Mediation and in the unceasing and eternal helps from God.

There is a rainbow canopy of etheric and luminous light

FIRST HEAVEN (Higher Astral World)

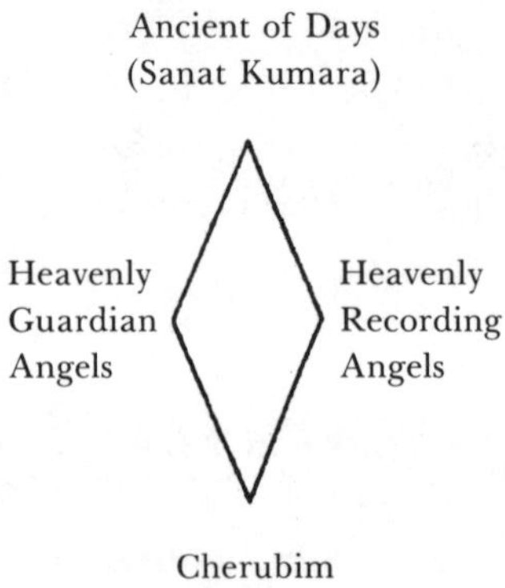

The Cherubim work to produce perfection in man's physical anatomy, and to stimulate the desire for perfection in physical objects. The Cherubim are the ambassadors who aid Jesus to perfect the ethical labors of man. The Cherubim overdwell the sacraments on the worship altars of men.

Sacraments of Heaven

Holy Presences	Buddha
Greater Cherubim	Bodhisattvas

*

Cloisters of Prayers

Ministering Angels	Saints

*

Pavilion of Light

Men in White Apparel (White Brothers)	Risen Dead Illuminati	Angels of the Dead

*

Higher Initiatory
Trials of the Night

Initiation through the Angels and the Great Immortals

Discerning the memory of past lives, past epochs, ages, aeons

Islands of Light

The range of the *Hall of Learning* extends from the Sacraments of Heaven through the Higher Initiatory Trials of the Night. The Hall of Learning enables one to qualify for night-serving, the use of ethic, and for blending former lives with the present.

encircling the earth. This canopy of pure light is called *the First Heaven*. The Saints, the true Prophets, the Holy Presences, the Elect and the Risen Dead dwell in this habitation of pure light.

The First Heaven is the higher astral world. Before the time of Jesus, the majority of men had access to only the lower astral chimeras and the First Heaven.

The First Heaven is the portal to the Kingdom of God. All who dwell in the First Heaven are communicable to the higher and exalted states of Heaven.

The action of the First Heaven begins on the fourth plane of the astral world. Here, one reads the akasic records and hears the audible sound or music of the Cosmos. On the fifth plane of the astral world, one unites with the Masters, the Saints, and begins his night instruction and night-ministry work. On the sixth plane of the astral world, he enters into the prophetic side of dreams. On the seventh plane of the astral world, one contacts the great Bodhisattvas who prepare him for initiation into the Second Heaven.

The First Heaven, as a holy circumference of rainbow-like light encircling the earth, is located outside the earth beyond the rim of the third abyss or the eighth sphere. The finer portion of light of the magnetic belt around the earth contains the four planes of the higher astral world or First Heaven. Jesus spoke of these four planes of the higher astral world as the "kingdom of heaven."

When Jesus said, "Heaven and earth shall pass away, but my words shall not pass away" (St. Matthew 24:35), He was referring to the higher astral world or First Heaven, the lower astral world and the physical earth.

Mastery of the astral sea assures one's penetration and research into Heaven. There are three Heavens. In the First Heaven one undergoes soul initiation; he reads his soul record. If his grace is sufficient, he is ready for initiation into higher degrees of light, and he begins to earn the right to research the reflected Spheres of Light, or Second Heaven, and to study their influence upon the minds and destinies of men.

The Spheres of Light constitute the Second Heaven. In the Second Heaven one is initiated into manifestation or Melchisedec powers. Here, he is instructed by the composite telepathy of the Masters or Great Immortals. The Melchisedec powers enable him to become a manifestor in right timing, and to de-manifest negations—with ethic.

The Third Heaven is the Kingdom of God, or the Archetypal Kingdom and the Realm of Light. The Lord Jesus now dwells in the Third Heaven surrounded by His twelve Disciples, the Most High Saints, and the Angels Celestial. The Christ Spirit centers His Light in the Realm of Archetypal Light of the Third Heaven.

Heaven is a vast community containing precincts and inhabitants of varied degrees of light. All form a body through which Jesus may do the Father's Will.

Heaven is more than a state of consciousness. Heaven is a World of Light correlating to the earth. Every earth created of God has a Heaven where dwell Holy Beings. The Beings inhabiting Heaven are Mediative Beings, Angelic Beings, Omniscient Beings, and Holy Presences of Light.

In Revelation 15:2, "the beast" is the Lower Astral World; "the sea of glass" is the Higher Astral World or First Heaven.

> *First-Heaven initiations are experienced in periodic, joyous, expanding, confirming states of Grace.*

Etheric color is basically supported by the higher pranic light. The etheric palette is stimulated by the vibrational or energy tones from the planets. The etheric colors are more distinct, less diffused than astral colors. Thus, geometrical designs and patterns formed by the coagulating aspect of ether may be seen by the seer. These designs are beautiful and inspiring at times. As one gradually moves into the finer levels of etheric consciousness, the akasic light in ether, using the finer essences of pranic light, produces scenes and designs of beauty and divinity. One begins to experience these divine visions in the First Heaven.

The soul seeks to write the history of Heaven upon the mind. The soul speaks of Heaven and its territories, its various parts and communities. When the disciple understands what the soul has to say concerning Heaven, he becomes a whole person in earth.

When one has received Protection-Grace in former lives, he is aware of Heaven, and he enlarges the vision of Heaven for other men.

The disciple should be patient with those who know not Heaven or who believe not on Heaven. Knowledge of Heaven must be earned. After having earned the knowledge of Heaven, one uses this grace to be a peacemaker, a joy-bringer.

One who speaks of Heaven is considered a cornerstone, a pillar for God. Only the love-tongue can speak of Heaven and be heard. The soul would teach the disciple to be more loving, that he may speak freely of the reality of Heaven.

The regular observance of the soul-faculty practices is as a yeast fermenting the unmanifested abilities within the heart, mind and soul. When the disciple uses the soul-faculty practices, he becomes a soul-linguist; he interprets Heaven through his thoughts, his words and his actions. His words and his works are laden with the oil of healing and anointing.

The disciple prays reverently to receive strength, guidance and wisdom from the Angels, the Presences and the mighty Beings of Heaven. He prays to become worthy to hear the Word as spoken through the Holy Ghost.

Through mantramic speaking, the clamors of the vibratory hum are softened. The karmic imprints upon the lesser etheric body are dissolved, and Heavenly telepathy is quickened.

When fasting is observed, the soul-faculties are magnified; the senses become the servants rather than the masters of the mind and heart. The powers of observation, memory-retention and pure perception become a charitable logos colored by pure feeling and thought.

The disciple contemplates to synthesize the manifested with the unmanifested. Through contemplation he stirs and extends the range of his three mental atoms. Through contemplation the power to image and the power to manifest become as one.

The disciple meditates to coordinate his emotions, thoughts and soul-faculties with the Inner Light of his soul and with the Eternal Spirit dwelling within the Higher Self.

May the holy dew of Heaven
Anoint and heal the sick.
And may the mighty manifestation of the Lord Jesus
Speak unto those who follow Him.
And let all who would follow Him
Remember that the Lord Jesus said,
"I am the door."

THE HALL OF LEARNING

All initiation in the Hall of Learning is preparation to serve as a night healer and a daytime initiate. Dream research of the night is under the supervision of the great Masters and pure gurus in the Hall of Learning. All spiritual aspirants are initiated in this precinct of Heaven, that they may render a knowing service in the world.

The Hall of Learning is located in the seventh plane of the First Heaven.

The inner schools most frequently visited and active are situated in the First Heaven. These are called the Hall of Learning. One experiences this instruction first during sleep in night-flight and in dreams. Also, one can be instructed telepathically by the Masters working with these schools during the day. If the student-initiate has mastered the art of daytime release of his etheric body, he can receive instruction during the day from the inner schools. This more often occurs after the student has mastered going *inside* in his meditation practices.

In the initiatory school of the night, one meets the Master in total freedom. In the daytime appearance of the Master, he can be seen only by an initiate who has known in many lives the technique of *seeing*. However, he can be *felt* with a form of ultra-sensitivity by one who is being prepared for the higher aspects of initiation. He can also be heard by such students with a form of *clear inner hearing*.

Dreams are progressive. Through night instruction in the Hall of Learning, one learns to extend the time limit of dreaming. Through spiritual understanding in evolvement, he learns how to use the releasing projectile, or the *silver cord*, within his etheric body so that he may remain longer in the tutelage precincts of the night university.

In initiatory dreams, all initiates must research the physical body, the etheric bodies, the astral and emotional body and the mental body. The great Overlords of instruction meet their proteges in the Hall of Learning or the universities of the night. Here, in laboratories of superconscious planes, their proteges are taught as to the functions of all bodies. All initiates having such night instruction become more aware of their need to learn of the processes of all bodies so that they may render a service on all planes.

In the First Heaven, where the Hall of Learning is centered, initiates attend night classes during dreams and sleep. These classes are overdirected by the Masters, the Illuminati, the Master Saints or Bodhisattvas, and also by some who while living in the earth are invisible helpers or instructors of the night.

In the Hall of Learning may be found everything that man must know and experience in the earth concerning his body and the use of his body. In the Hall of Learning in the present time a greater activity is occurring. This pertains to science and man as he will be in the earth in the next 3,000 years. Everything that man is to manifest in the physical world is first worked upon and blueprinted in the Hall of Learning before it is utilized by the ingenuity of responsive, competent minds in the earth.

Many initiates on recalling their dreams put into action their diagrams and blueprints received while being instructed in the laboratories of the Hall of Learning.

All ideas of the night initiation eventually drift downward into the receptive, higher intelligible minds and souls in the earth. To have cognizance of such dreams and night instructions is a confirmation that one is indeed fulfilling his true task in the physical world.

When one enters the Hall of Learning for instruction, he becomes receptive to his instruction through the color of the rays of night. His etheric body and his higher emotions become receptacles of cosmos light, vibration and color. From this he becomes a daytime channel and presence of healing love.

In the Hall of Learning the initiate crosses the threshold into the mysteries of Heaven. The majority of the initiate's experiences in the Hall of Learning occur during sleep. However, each initiatory experience of the night invariably correlates to some action in the waking state.

All who enter into the Hall of Learning come under the direction of the Cherubim Angels, the custodians of the symbolic codes of the First Heaven. The *freshet poems* received by the initiate are under the direction of the Cherubim Angels. Music, poetry and art are made more profound when one is instructed by the Cherubim Angels.

The emotions and thoughts of the quickened dead who are being prepared for birth are filled with the delight of basking within the light showered upon them from Heaven. The reading of their souls' records of past lives, revealing to them the justice of God's Equation; the Light of Heaven which is reflected downward upon their everlasting bodies, and the instruction received in the Halls of Learning—all prepare the quickened dead to return to physical birth.

In the Hall of Learning, men study minute subjects, such as the cells, during the time when the constellation rays focus their velocity upon the planet Mercury. This in turn is stepped down into physical laboratories of the world, that all of the minute or microscopic forms of life may be researched.

Initiates having Mercury elevated in their charts may be found at night in their dreams working under the direction of the greater cosmic presences. From such instruction and apprenticeship come men who have talent for the intricate and infinite, or patience with a side of life ordinarily unobserved by other men.

In one portion of the Hall of Learning there is a mighty, transparent domelike roof or covering. The material forming this transparent ceiling is supercharged with pulsating energy that acts as a vibrational receptor and distributor of the great planetary and constellation rays of Hierarchy beamed from the outer constellations in space. These Hierarchy rays beam into the vibrancy mental coils of researchers of the night, giving an informing action. These great energy rays determine what men are to learn in particular times in the cosmic laboratories in the Hall of Learning.

In the Hall of Learning there is one place where nothing but love is taught—how to use love so that God can use you as a love-vehicle. When you use love blindly, the Love of God cannot use you.

In the school of love, one learns that love heals and prospers everything. The first thing one learns is to release everyone he loves to the Love of God. Then the Love-ministering can begin.

Love as an energy works through the heart and through the ear. When one becomes a sounding-board for God through love, his voice becomes a healer to those needing, yet not knowing they need, love.

Love in the heart is the great timer for the body, for the soul and the mind. One is taught in the Hall of Learning that he must commune with the Love-Waves of God. Through union with the Love of God, the mind is filled with a supramental love, and he is beyond thought of taking love to satisfy the senses.

Through the feeling of love in the inner organ of the heart, where the flame of love houses the Love of God, one emanates love and can know no estrangement from the Love of God.

There is a peculiar clairaudience that enters the ear of love's sensitivity that must be transposed through speech, or an instrument of God's Love suffers save he can speak with love. This is the healer who is attuned to the Love-Vibration of God in the heart, in the mind and in the ear. This is a Saint and a Sage on earth who loves as a vibrating-center for God.

My heart overflows with the Bountifulness of God. I shall see His Provision for every man. Let my love for God trust in His mighty Provision.

THE SAINTS

As men make paths for their physical feet, so corridors are made in the Higher Worlds for the coming and going in their etheric bodies. As men build houses, records are made of their labors. As men make tunes or songs or compose symphonies, Heaven's Light is extended wider into earth, and all on earth share the Light of Heaven.

Each healing contains something of Heaven. Man without Heaven is a darkened island unto himself.

Response to Light is innate, natural. The body, the cells, the blood, the bones of man having been envisioned in Heaven, it is natural as to the order of creation that the organs, the emotions and thoughts of man should *respond* to Light. The consistency of all organisms being of Light, it is in accord with God's Equation that all men respond, in some manner, to the Light.

One cannot receive the Kingdom of Heaven while yet adhering to the lusts of the world. The rules, the ethic and the law provide the disciple with a way through which he may sustain his spiritual gifts. When the disciple receives a change of heart, he attains a higher degree in his evolvement. He fortifies his evolvement by working willingly in the small tasks; for he must first prove himself worthy in the least things, that he may prove worthy to receive the greater responsibles.

An accumulated number of martyred lives produce the Saint. The true Saints remain within the First Heaven and rarely re-embody. Those who have yet to become Saints through the powers of the Higher Worlds, rather than through ecclesiastical authority, are called "worthy ones" in the Higher Worlds. Such worthy persons return to the earth when the souls of men are endangered by the Antichrist actions in the world, and fulfill their works of Light despite persecution and martyrdom.

The *worthy ones* are saints-to-be. The Anointed Saints are Saints living on earth and in Heaven; their supernatural precincts are in the First Heaven. The Heavenly-ordained Saints are Saints whose supernatural cloisters are in proximity to the heart of the Lord Jesus in the Third Heaven.

When one is initiated into the supernatural precincts of the Saints in the First Heaven, he first comes under the direction of the Anointed Saints. As the initiate evolves in his selfless works, he enters into an extended degree of light whereby he comes under the direction of the Heavenly-ordained Saints. Under their tutelage he learns of the miracles of Heaven showered upon men of faith, as the Heavenly-ordained Saints have the power to de-manifest malignant substances and to transubstantiate them into purity and health. They also have the power to rain upon men the holy-bread tokens, provisions, special gifts and blessings.

The Mother of the World is a Divine Being representing the Feminine Principle. All women come under Her protection and guidance. The one called Mary, the Mother of Jesus, and the one called Kuan Yen of ancient China, work directly with the Mother of the World. The women Saints of Heaven and also the women adepts, called *Fountains* in the Higher Worlds, work with the Mother of the World.

The Mother of Jesus works with the mothers of the world. She does not re-embody or return to the physical world. Saint Anne is the coordinator for the women Saints; she re-embodies or returns to the earth in timing to world need.

He who stands on the pinnacle of persecution and is the target for crucifixion is working with the Saints and will someday become a true and holy Saint.

In the First Heaven there may be seen a network of roses which symbolizes a Saints' cluster and network of sainted souls. Their holy rose-fragrances envelop one who is to be a

Saint. The oil of roses is the oil of sanctification. He who works with the Saints lives within this nectar.

The Saints are non-passional and all blessing. He who walks under the Saints forgives all, censors none. On the chain of virtue ideas, one receives the downpouring of the holy roses. On the chain of selfless devotion and love association in the heart, one receives the attracting powers of drawing to him divine companions in the walk of Light. These companions are in both the physical world and in Heaven. In troublous initiatory times, one must lean more on the heavenly companions than on physical divine-companions in the world.

The initiate works to achieve a simultaneous alignment with the Saints, the Bodhisattvas, the Illuminati and the Masters. The Saints represent the Love and Life Currents; the Masters, the Will and Mind Currents; and the Illuminati, the Light and Etheric Currents. All work within the Sound Currents of the Word to teach, to heal and to anoint the initiate. One begins by uniting with *one* of these Currents.

BUDDHA AND THE BODHISATTVAS

> *The one known as Buddha, who lived approximately 500 years before Jesus, now dwells in the First Heaven, called Nirvana in the Eastern teachings. Working with the Lord Buddha are many Bodhisattvas or Sacred Men.*

The First Heaven is the high side of Eden. Buddha is the center in the First Heaven, and the Bodhisattvas are his rosary or necklace companions.

The Bodhisattva Light dwells in the higher Deva Planes of the astral. Buddha functions in the inner worlds at this time through the Great Shining-Spirit Deva Light. In the

Christian theology, the higher Deva Planes are called *Heaven*. In true terminology, the Deva Planes are, in reality, *the First Heaven*.

Buddha, living before Jesus, came to prepare man for the Christ-Spirit Avatar Light of Jesus. Buddha united the mankind-ego with the Psyche *mind-functioning* of the soul. In Buddha's own enlightenment, he opened ten of his chakras, with emphasis upon the third chakra. His exploration-initiations gave him contact with, and command of, the third-chakra gate of all mankind. Buddha, as a directing, mind-illumining Being, now dwells centered in the First Heaven, where all third-chakra action is mastered. All Bodhisattva-action under Buddha functions through the third chakra.

Men are presently working to polarize the Lunar-Avatar light with the Solar-Avatar light. The Lunar powers of Buddha are open to one being initiated through the third chakra lunar-light. This lunar light unites one with the First Heaven, where Buddha is in command as the central Avatar.

> *The Bodhisattvas in the First Heaven sound the tone of Om, thereby helping to sustain the reflected archetypes overdwelling the First Heaven. The Christ-Spirit sustains the inner Tone of the Greater Archetypes in the Third Heaven or Kingdom of God.*

Auric Greatness Powers

> *Lead me, O Blessed Presences, to the healing pools where my fevers of uncertainty are stilled.*

The vibration of elevated auric power, when pure, heals. Auric tones known to the intuitive soul can be intuited by an expanded spiritual listening heart and mind. When one stands in proximity to a soul having a pure condensed auric power, he is flooded with spiritual power. The inter-exchange of auric soul harmony overflows into one who seeks to commune with auric greatness. The auric greatness power is a mantle power. Elijah left his auric greatness power as a mantle to Elisha.

A phenomenon of high spiritual quality is transmitted from the auric encirclement of the Masters and high Bodhisattva-gurus to their chosen proteges. To stand in the proximity of one who has received auric transmission from a Master is to receive healing and rearrangement of one's most heavy karma.

In the *veldt* or *bliss* vicinities of the highest astral world, or the First Heaven, are Bodhisattva-Gurus of auric power to be received into the auric caves, or sacred openings in the auric fields of great mystics, sages and prophet-initiates. These transmitted sound waves keep the proteges of Bodhisattva Saint-Masters alive to spiritual power. Received as a transcending light, they fall upon the need of the reverent, accepting mind and heart, giving to an aspirant of the spiritual life the power to pass on to those who are in need a way of hope, of light. These auric-greatness happenings come as radiant visions, as inner hearings, as revelations.

To overpower evil, to exorcise, to produce pure holy phenomena, one must pray to open the sacred caves of his aura. These sacred openings are vortices of cosmic energy waiting to be freed into the need of man. Anyone having the sacred caves or cavity vortices opened in his auric field is energized by a tireless flow of resurging life vitality. He is also a master of timing. Close to Christ are such ones. They are always turned in their faces toward Him, knowing that He, the Christ, used His auric greatness to stem the wicked and unruly forces in the world, to bring peace to the unpeaceful, righteousness to the persecuted, and joy to the hopeful.

Only love can overcome the barricades built by selfishness. Let me love.

THE ILLUMINATI

The Illuminati are the genius Risen Dead whose instruction enters into your telepathic stream of consciousness to lead you to recover your soul-talents from past lives and to produce new talents in this life.

Through the Door of Jesus, one unites with the Great Illuminati who work with the twelve creative, prototypal, dimensional states of consciousness. The Illuminati are in alignment with the Immortal Masters who work with the twelve archetypal dimensions of the mind; the Divine Bodhisattvas and their saintly guru associates; the immortalizing Angel companies and companions; the shining and holy Devas; the planetary and Hierarchy system.

One of the higher orders of the Risen Dead is that of the Illuminati. The Illuminati are creative titans. They are advanced men who lived in the world and produced mighty works of art, which taught men of the eternal realities and inspired them to live a spiritual life. After death, the Illuminati work with the Men in White Apparel to instruct the Risen Dead and the quickened dead who are destined to create in coming lives. The Illuminati also work with the living who are ripe to express their creative potentials. The greater art projects, such as immortal sculptures, music, murals, literature, and architectural structures are inspired by the telepathies of the Illuminati.

The Illuminati are in direct and frequent communication with the living. If one living has the knowledge of the Illuminati's overdirection and supervision, he learns that he receives his greater inspiration during certain times when the higher planetary energies are conducive to art, beauty, design and perfect form, for the Illuminati use the higher degrees of light extracted from the energies of the planets to inspire those who would dedicate their genius to God.

Eternal Grace is grace one has earned in other eternity systems previous to birth in this earth. When one entered this eternity system, Eternal Grace was sealed into the central point of the Diamond Medallion around the Higher Self. Eternal Grace is watched over by one's Recording Angel. Grace from one's actions in other eternities is spiritual power. When one makes alignment with Eternal Grace, he brings to the world a teaching of man's "everlasting life" as of the Eternals, rather than that which pertains only to this eternity. He is able to communicate to men the imperishable knowledge of eternal or everlasting life. He belongs to that company of Elect known as the Illuminati.

The Angels are the first life-companions of man. All Omnipresence Beings assisting man that he might attain awareness of God exist within the Angelic Kingdoms. Associated with the Angels and sharing their life and light vitalities are the Saints, the Illuminati and the White Brothers. In direct contact with these are the Masters and the Archangels. And always with the Archangels are the Shining Spirits or the Greater Devas.

In the Sixth Light-Stream work, one encounters the Ishta gates—that is, the most close Omnipresent Beings to his Illuminati Grace. These flow into his soul-powers encased in an etheric circle or mandala with four gates. At each gate is seated a Maha Deva which God, Life and Fate have assigned to him over the many aeons and ages.

The world of the Illuminati is a golden city of light. Any person who has any contributing knowledge to leave to the world is in the golden city of the Illuminati.

If the poet is too ecstatic in the Cherubim Realm, his desire to remain in this blisslike state may detain him and upset the rhythm of the after-death progression. His Personal Angels watch over him and shepherd him upward toward the world of the Risen Dead, that he might begin his work in Heaven, where as one of the Risen Dead he may work with the Illuminati. The Illuminati, right-hand companions of the Men in White Apparel, have achieved immortality on earth. Their work is to inspire the poets, the musicians, the writers and the artists of the world.

The true Rose Cross is a Fraternity of Light existing only in Heaven. The Brothers of the Rose Cross are in reality the Illuminati, who work at midnight to maintain balance within the four gates or points of the earth's etheric compass. This is called *the fulcrum work.* The negations, the darkened passions of men of the earth, and the subtle challengers of the lower astral worlds, at the time of midnight send forth a volcanic-like upsurge of the dark to challenge the power of the Light. Each night at midnight victory is made over the dark. In the morning, men arise with peace, inwardly remembering that darkness is but an instrument, and that Light is eternal and indestructible.

During very special world-crisis periods in which civilization stands upon the threshold of new eras in creation and good, the Illuminati and the Saints fuse their works. The Illuminati-Initiate powers and the Saint-Initiate powers are made omnipresent among men. The Illuminati-Initiate takes mercy into his vocational fiat. The Saint-Initiate becomes creator as well as anointer. Thus, in this period of merging Initiatory powers, a greater expansion of the World-Soul Medallion is made possible. Light enters into the cultures, and the humanities are blessed.

The specific goal for the advanced Initiate is to make equal his power as an Illuminati with the providence of his Sainthood mercy. In this period of renewal in Light, the Eternal Diamond Medallion of the Christ and the Tone of the Word of God are made manifest through men of pure ideals, generating and expressing selfless good.

When a person unites his Illuminati Grace with his Mercy or Saint Grace, he becomes a Cosmos-Disciple.

THE RISEN DEAD

The Risen Dead are a handful of God who dwell close to the Saintly Cloisters in the First Heaven.

The various phases of instruction in the lighted corridors of the Risen Dead are wafted downward into the world of the living. Persons in the physical world who have spiritual and philosophical inclinations are influenced by this instruction in the daytime hours and sometimes in the dreams of the night.

When the Risen Dead have attained command of the magnification of their thought processes as expressed in Heaven, they are taken to the Hall of Records, where they read the grace-records of their former lives and the grace-records of the lives of persons they have known intimately in the just previous life. The reading of the grace-records continues for a period of time so that the Risen Dead, through extended logic, might incorporate into their minds the Spirit-rationale. This is the rationale beyond philosophy or retrospection. The Spirit-rationale stems from the Ultimates preordained from the beginning of the world.

After the reading of their records, the Risen Dead receive a holy insulation. Thereafter, it is impossible for the living to "bring up" or to wilfully communicate with the Risen Dead. It is possible to "bring up" the unrisen dead; it is also possible to intrude upon the quickened dead. However, no occult or psychic power can touch or reach the insulated Risen Dead.

At regular intervals, the Risen Dead detach themselves from the Risen Dead activity and undergo a respite-interim, which is a form of visual sleep. The closer one of the Risen

Dead comes to rebirth, the more frequently he falls into a respite-interim. The Risen Dead must undergo respite-interims previous to birth to enable them to prepare for the coming life on earth. In this, they do not review their past, but they preview the coming life.

In the lighted corridors, the Risen Dead use the soul-faculties rather than the senses. The Risen Dead respond to the visible and audible pictorialization in the respite-interims in a different manner than the unrisen dead. The Risen Dead experience sound and light as intelligence. Their emotions, free to function in the First Heaven, record the heavenly music as ecstatic love.

In the cosmic seasons, the Risen Dead and the Recording Angels work closely with the living. The Recording Angels reveal certain cosmic truths to the spiritually inclined. The Risen Dead are fortified by the major spiritual impulses flowing into the earth; they are enabled in these periods to send their telepathies freely to the living. The Risen Dead send their assurances of the immortal life. Their heavenly love sent to their beloved ones in the world strengthens the living that they may bear their trials, and blesses them—giving them a supercourage to surmount and to overcome the rugged trials in the physical world.

The telepathies of the Risen Dead enter into the minds and hearts and affairs of the living in right timing. Some who live in the world intuit this knowledge and respond to it with awareness. Others who live in the world, being less certain of a life after death, interpret this help as a form of startling coincidence, or as some happy and special "good luck" for them. Regardless of the belief of the living, the Risen Dead come forth at intervals and give help to their loved ones, that their life on earth might be more hopeful, more blessed.

A special emphasis on repentance is placed upon both the living and the dead during the Autumnal Equinox. The living and the dead are offered the opportunity through the

Archangel Michael to face their debts and to repent, or to atone. In the Autumnal Equinox, the quickened dead awaiting birth in the pre-birth matrix read their former-life records. Their Guardian Angels give them the choice of three paths: (1) the path of family and blood-tie discipline; (2) the path of struggle and self-attainment; or (3) the path of serving the human cause.

During the Autumnal Equinox, the Risen Dead in the pre-birth matrix are enabled through the Archangel Michael to read the grace-records of the martyrs, the saints and the prophets, and to take upon themselves the mantle of grace so that they will not be led astray from their purpose in the world.

There is but one path for the Risen Dead who are reborn to the physical world—the path of sacrifice and service. Even as Jeanne d'Arc, under the direction of the Archangel Michael, chose the harder way before coming to birth, so do all saints and martyrs enter the world without the expectation of a life of luxury and ease. When the Risen Dead live in the world, they choose simple and humble environments, for it is necessary that the spiritual person be free from materialistic hindrances and confinements.

The Risen Dead, previous to rebirth, are instructed as to the Second Heaven where dwell the Great Immortals, who are part of the Elect. In this instruction, the Risen Dead learn of the Planetary Angels, of the Host or Hierarchy. During this period, they incorporate into their everlasting bodies the finer prisms of spiritual light, which insulate them and prepare them for future initiatory experiences in the world.

The Risen Dead are familiar with the power and momentum of prayer. They behold the prayers of men, the angelic helps accompanying prayers, and the differentiation of the varied degrees of prayers. The Risen Dead, when sympathetic to some earth need, work with the Saints in the transmutation of prayer. The Saints transmute all pure prayers upward into the Garment of the Lord Jesus. Jesus and His Disciples dwell in the Third Heaven or Kingdom of God. All who pray to the Father in the Name of Jesus receive bountifully.

SPIRITUAL TIDES

I worship God Who inspires man to create all reverent objects.

When the soul of man becomes diamond-like, the facets of his soul become a prism-hourglass for the Christ; man's timing and his grace become as one. He responds to the spiritual tides of Heaven—and all who stand near him come under the mantle of his grace.

All persons have grace, but not all are free to use the assets of their grace. To free one's grace, one should revere God and make a covenant to qualify himself to fulfill the Law of Timing. He begins by seeing God's Equation working within all events.

If one fulfills his smallest and most menial tasks with diligence and with joy; if he fulfills his promises to himself and to others—he will come into timing and he will reap the greater benefits of his grace. He will become not only a diligent and trustworthy person, but he will come under the guidance and supervision of the Holy Presences of Heaven, for he will have made himself a worthy timekeeper for Heaven.

When one has opened his spiritual grace, he reverently anticipates the spiritual tides. Not all persons respond to the same spiritual tides. Some feel the need to commune with God during the spiritual tide in the early morning; some in the noon day; some at dusk; and an increasing number of persons unite themselves with the spiritual tide occurring at 11:00 P.M. The spiritual tide at 11:00 P.M. is the period in which earth-initiates receive a unique rejuvenation and instruction.

Those who are completely engrossed in their physical activities of the day are unaware of the spiritual tides. However, while sleeping, their souls respond in some degree to the spiritual tides of the night.

Anyone who fulfills a continuity in spiritual serving intuitively responds to each spiritual tide in its sequence. If a person has the grace to respond to all of the tides, he has a fluidic cognizance and a range of action beyond the average individual. The resuscitating helps of these spiritual tides reinforce his labors and give a spiritual vigor to his serving.

The spiritual tide occurring at sunrise is a tide of animation and quickening. Those who respond to the spiritual tide at sunrise often become ambassadors for the light of Heaven; their souls' prompting stirs them to awake and unite with all waking and stirring life of the earth. One whose timing is alerted to this hour experiences within himself a mediative current of exquisite peace and upliftment. He unites himself with Heaven's design, and he sees men arising from their sleep as creators in a fertile world.

One who regularly awakens *after* the spiritual tide of sunrise may unite his thoughts with the instruction received from his soul during the night's sleep.

Ten o'clock in the morning is a propitious time for dedication, because in this hour the soul reminds one that he is a creator. The spiritual tide of 10:00 A.M. stirs the intellect. The tasks of the day are clarified, and one may coordinate his emotions with a hidden strength to meet the day's demands.

The spiritual tide at twelve noon enables one to contemplate the magnanimities of Heaven. At noon, man's wonder-consciousness and his naivete are at their height; there is a hushed moment in which he may be free of harsh judgments toward his fellow man. He may use the meridian powers of the noon sun to conjoin his prayers with the prayers of the Mother of the World. At the exact moment of noon each day the Mother of the World places her garment of forgiveness over the erring ways of men. All the powers of forgiving in Heaven are fused; the Angels, the Saints and the Holy Presences of Heaven experience a rapture of union with eternal good—and they envision man as he is to be.

The lesser etheric body ceases to respond to the influence of the physical sun at 3:30 in the afternoon and begins to respond to the lunar rays. This transition from the solar to the lunar rays affects the emotions of man and enables the soul to come closer to his mind.

From 3:30 P.M. to the time of the dusk spiritual tide one moves closer to his conscience. If he inwardly feels that he has fallen short of his creative aims, he is weary and discontented. His egotism is at its lowest ebb, and thus he is more likely to respond to the spiritual tide at dusk. When one's thoughts are retrospective, and he reflects upon his actions and motives of the daytime hours, he has the opportunity to gain a deeper perspective and insight into his motives, for in the dusk time his individual will is less forceful. If one is highly evolved, he feels a humility and a self-inadequacy. The demotion of self makes it possible for him to turn to the greater help received from the Lord Jesus during the spiritual tide at dusk. In the violet dusk of the sunset hour, the Lord Jesus comes closer to men; and those who will may see themselves through the perspective of Heaven. At sunset or dusk the curtain rises upon the spiritual life of man, and the Lord Jesus, the Sovereign of the human spirit, moves tenderly and closely into the open hearts of those who lift up their prayers at day's ending.

In the world there are certain anointed ones who have given themselves over to the very special mission of healing the wounds inflicted by charlatanism and exploitation. These anointed ones, understanding the law of transposing one current into another, work to overcome the psychical tumults which are rampant at 9:00 P.M.

Each night at eleven o'clock the molten sea within the earth, the axis of the earth, and the innermost point of the Kingdom of God are united. This heavenly conjoining is called *the fulcrum hour.* A mighty upsurge of spiritual power is generated in Heaven and in earth—touching all living organisms, sentient beings, and degrees of consciousness. The sum total of the evil consummated in the day just past is mitigated. Thus, evil can never become victor in the world. Evil will ultimately be replaced with good.

Those who respond to the eleven o'clock spiritual tide work with the momentum of the fulcrum action. If they have the power to witness the fulcrum action, they see that the earth is enveloped by a mighty network or medallion, and that at each point of the medallion there stands a spiritual sentinel or Holy Presence. From 11:55 until twelve midnight the highest point of illuminative light exists in Heaven and in earth—and the upsurge of the fulcrum action culminates in a world-purging, cleansing, healing.

There is a tryst tide before the early dawn of each rising day. This occurs at 3:00 A.M. Advanced Disciples, Telepathic Disciples and Cosmos Disciples meet in the etheric realms where they work for the peace of the nations and continents.

My soul has a memory of the Rituals of the True Sacraments. I respond to God when my soul is quickened through love.

MELCHIZEDEK and Manifestation Angels

Melchizedek works with the Great Immortals within the Spheres of Light to teach the initiate how to overcome the four elements of the Earth (fire, earth, air and water), that the initiate may obtain the power of Manifestation and De-Manifestation.

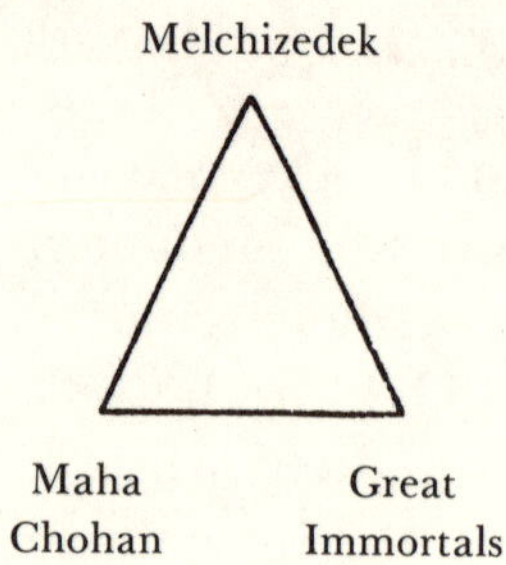

GREAT IMMORTALS and Seraphim Angels

When one has attained the higher degrees of initiation within the Spheres of Light, he will receive instruction through daytime telepathy from the Great Immortals. The Angels of Seraphim or the Planetary Angels are called by some the Lords of Mind. In night and day initiatory experience, the mighty, rhythmic Angels work directly with the Great Immortals to interpret the Archetypal truths to purely evolved persons.

MAHA CHOHAN and Recording Angels

In night or dream instruction, the pure disciple is initiated in the Spheres of Light through the Maha Chohan. From the Maha Chohan, the initiate learns of the spiritual governings affecting the governments of the Earth and the laws of the Earth.

SECOND HEAVEN: SPHERES OF LIGHT

14

THE SECOND HEAVEN: THE SPHERES OF LIGHT

Men having earned the power of traveling in the Higher Spheres release their etheric envelopes or bodies and carry out the research onto the Sphere of Light of that which is sought and to be used in Earth. They travel through the release of the silvery cord from the magnetic and the electric upward through the astral world, the world of mind, and into the World of Light.

The Hall of Wisdom

The Hall of Wisdom, colored in peachlike tones with violet hues, is seated above a massive portal of stars. Directly over the central column, there is a medallion containing a triad. Within the medallion and the triad stands the symbol of Man. The scroll beneath the symbol reads: "Invincible, the Eye, the Will, the Love. The Creator hath erected this dwelling. All who walk here are Immortal. Men come as eagles from the West, as doves from the South."

In the School of Shamballa, or the *Hall of Wisdom*, initiation into hierarchy powers is experienced by the initiate. Here, in the planes of the Second Heaven, called *the Spheres of Light,* initiates become Teachers, Teachers become Gurus, Gurus become Masters, Masters become Bodhisattvas.

During sleep, the initiate who has reached a certain stage of evolvement often dreams he is receiving instruction in a classroom. In some instances he recognizes those who live with him in the world. And often he is aware of a dear familiarity uniting him with the speaker or teacher in the classroom. These classrooms are, in reality, the corridors of the Spheres of Light or Hall of Wisdom.

All symphonic compositions are inspired by the true music within the Spheres of Light. The *Music of the Spheres* consists of the tone within each planet accompanying the earth. These tones are captured into the core of the earth, and they resound within the Spheres of Light. All initiates of music, or composers, penetrate these tones within the Spheres of Light; such persons bring music to the earth to overcome the karmas of men.

Music from the Spheres of Light is a heavenly combination of sound to disburse and dissolve the karma of individuals and nations. Conditions and society may be transformed through great music. This music in no manner compares with the atavistic or primitive music correlating to the lower regions of the astral world.

The rhythms of night-flight come with practice and with heart-searching love. The corridors in the Hall of Wisdom place men in Light, enabling them to work on earth as future leaders. The mission of each one as disciple is to first seek the ruby tone; then the tone as fire in green; then the lilac point in Light; then the sapphire as of the blue; then as of the amber fire; then of the tone of indigo; of pearl; blending in the rays of the Master's Light; blending in the ray of color; of sound; of lifting in consciousness; lifting deep into the color; seeing only the ultimate within the violet Light, the forming pictures above the mist; the revealing within the mysteries.

The Hierarchs are Great Beings who overdwell the zodiacal constellations around the Sun and the planets of this earth system. The Holy City, called Shamballa, is in reality in the Spheres of Light, where the Hierarchy send their Rays. These Rays become Light Streams; the Lord Jesus transposes the Rays into Light Streams for men of the earth.

The Rishis are Elect beings with powers correlating to Hierarchy. They originally came as Source souls from the eternity-system cluster of the Great Bear. The Rishis bring the *Laws* for interpretation of karma. Their present station is the Second Heaven with the Logoi or Planetary Archangels who overdirect the reflected spheres of the planets.

There are seven great Rishis who gave the *Vedas*, the *Upanishads*, and the Abraham thread within the Judean teachings. The Abraham thread also includes the Islam reflection of the Rishi. All teachings originate with Rishi teaching. The Egyptians were heirs to the Rishis. The Christ, as the only begotten Son of God, is beyond the Rishi. In the Christ, one coordinates all teachings and laws with love.

The Hall of Wisdom school in the white light of akasia unites and flows into the seven initiatory polarity points in the Earth where the overdwelling Master uses the earth-atoms in his higher etheric body, appearing to the key initiate residing in the polarity point.

Jesus, Master of the Masters, Lord under the Lord of All, needs a worship-vortice through which He may work with the masses. He also needs a body or nucleus of specialized souls to work with individuality. In all worship-bodies where the Masters synchronize their labors uniquely with Jesus, a very wonderful message can be given to the world.

It is needful to worship; it is also needful to work toward higher initiation that one may truly serve in the world. All initiates will remain in the lesser instruction until they understand that Jesus is the Archstone upholding the building of spirituality. The Masters since Jesus can do a mighty work; the greater things "men are to do" are made possible through Synchronization and Mediation.

Some have seen visions of the Lord Jesus in telepathic light, for He is the Great Teacher and World Saviour, the true Mediator for all who are in the Light. Alignment to the Masters or Great Immortals within the Higher Worlds makes it possible for one to enter into telepathic communion with the words of the Lord Jesus. This only occurs when there is some unusual state of illumination.

An *Island of Light* is a plateau of Light built from the prayers of men. Upon these plateaus or Islands of Light, the Great Immortals await those who rise in the night to the Spheres of Light or Hall of Wisdom.

Many seers and mystics on Earth have known of these Islands of Light but have not known their true relationship and interpretation as to the Spiritual Spheres. The Spiritual Spheres of Light are not of the astral world. The Island of Light compares not to the astral world but of itself gives men Light, aid, and help out of the Master's great Aura on which He receives from the Aura of Christ onto the Hierarch's work in Earth.

> *He who dwelleth within the Master's Aura carrieth himself close to the Spheres of Light, but toucheth the higher regions of the astral world. He who dwelleth in the Master's Telepathy toucheth the Spheres of Light and beginneth his work in the Hall of Learning. He who traveleth with the Aura, Light, Love and Telepathy of the Master findeth himself in the Spheres of Light radiating to all the Masters in the Earth and the Masters in the Spiritual Spheres. He walketh freely in the*

Hall of Wisdom. He is attracted there by his needs according to his own soul-karma and the karma of those with whom he mingleth.

THE HALL OF ARCHIVES

The Scrolls are read in Light when the eyes have gazed over many lives unto a distant place for which the yearning is so intense that the Sacred Atom within the heart is loosely set, and nothing of the sensual discoloreth the heart. When the Scrolls and the Archives are read, the moving panorama of life is brought forth, and man beginneth to relate himself to the rational part of God and His Word in earth.

The Hall of Archives is in the Hall of Wisdom. The walls are filled with scrolls and outlines of happy figurines, joyous Cherubim, and a panorama of moving rhythm, floating harmony, joy and musics.

True spiritual beings live in a continuing state of action and joy. Those who inhabit the Heavenly Spheres commune with other Heavenly Spheres. They commune with the vowel-tone of love and with the knowing in love. Etheric beings who have never inhabited physical bodies commune through waves of joy and light. They carry their mission to earth and to the human life-waves through the somber tones of mercy mingling with love. This enables them to travel in the rhythms of light through the octaves of sound set up within the fourth plane of the astral world.

The peace of the Inner Spheres is not a static peace. It is an active, moving, living light, and reflects itself into the minds, hearts, and onto the tongues of men. The whole universal voice of praise is light. And love worketh through light.

In the Archives of Light are recorded the struggles of men's souls, of their walk ever toward Light, of their reaching out toward the Good. There is also recorded the sayings and the workings of the dark.

In the Hall of Archives is an unending enactment or panorama pertaining to man's future state of fulfillment. This panorama is accompanied by the Cherubim Angels who work with man's music, industry and architecture. This panorama is perpetually sustained in the Archives of the Higher Worlds so that it may enter the stream of man's consciousness and inspire him to fulfill on earth what Heaven envisions for him.

In the great Archive libraries in the Spheres of Light, there is the record of everything written in Earth. This includes the time in which man took clay and filled his forms of clay with symbols out of the sky, the fire, the tree and water. Over the ages, men built their conic, their hieroglyphic, and their outstanding alphabets as to expression in vowel and sound. On this day men read, study, work, and refer to one another's ideas. The Archives record these works and these ideas. The library of the Archives also records the books to come into the Earth—the books which will bring to men a new thing in their consciousness, a new setting up of energies and ideas.

All great disciples leave their record in the Archives of Light which may be read by those going into them through Light and discernment. On the reading, they see the human karma and the spiritual balance through which those working in Light carry with them the cloak of martyrdom in Earth. The advancement of knowledge above that which man has in his consciousness, coming through any human being, brings always the penalty of karma, its disciplines and sorrows in the Earth.

The astral world is as a great serpent never ceasing to move. When the Spheres of Light have accomplished their work, the moving serpent of the astral world will become static or still. Men will use the powers of the astral world to build the thought forms of the future on which creation comes forth without flaw, perfected after the genius of instant timing, instant sending and receiving.

The Great Beings, the Elder Brothers, the Masters, the White Brothers, all live between and in the Spheres of Light and the Earth. Were it not for the Spheres of Light, the Masters would find no means through which to touch men of Earth. Were it not for the Spheres of Light, men could not travel into the areas in which the great reflections and pictures are revealed as of the Archives, akasia, and the things in the Earth which build out of Light.

The Spheres of Light are not affected by the planetary forces in their negations, but only in the greater impulses of the Hierarchs are these Spheres affected. The negations of the planets fall into the astral world as sound.

RITUAL AND THE SCROLLS OF LIGHT

> *One has to move beyond the 33 precincts of Jehovah in the lower astral regions, that he may reach the Second Heaven where dwells the true Archetypal formats for all religions in the Earth.*

Three thousand years before the time of the Lord Jesus, men detached themselves from the vitality of the Ritual of the Higher Worlds, preferring to use the magical procedures in ritual rather than the protective and disciplinary Ritual of the Higher Worlds. Thus, men began to worship idols. In Egypt, occult ritual became the decaying black magics. In Greece, Hierarchy became the myth-gods. In Rome men fell into the worship of the myth-gods and idols of other men. In Babylon and Assyria, men worshiped the idols of brass. The worship of Mammon, Baal and Moloch depraved men and isolated them from the Higher Worlds.

Three hundred years after the establishing of Christianity, the early Christian church established a system of ritual. This ritual succeeded in building a secular and formal ecclesiastical framework of worship, but repressed the former, pure Christian practices as used by the early Christians. Thus, from this period, men were depressed into materialistic worships and viewpoints. With the coming of the new ideas from the Higher Worlds, men once again will have the opportunity to correlate their rituals within religion and worship with the true Ritual of the Higher Worlds, and thus produce a spiritual renaissance within religion.

In the just previous life, many persons with agnostic hearts had been victims of oppressive religious ritual or bigotry. Such persons come into the world with an aversion to any sort of religion, particularly to a religion founded upon ritual.

In the next 3,000 years the true rituals of Heaven will purge away the crystallizations gathered from the physical concepts in religious ritual. The Age of Science will introduce new symbologies which will free the true rituals of the Heavenly Sacraments.

In the great Cathedral of the Sacraments, man-made ritual in its negation is overcome by Divine Ritual, of which music, shaping and creating, moves as rays and waves in tone and counterbalances the rigidity in ritual as of earth.

The pure Ritual of the Higher Worlds now being transposed and sent to the earth will find in its beginning only the blind side in men, and resistance.

The Great Immortals or Masters are in command of the Divine Ritual in the Higher Worlds. Through the use of the Hierarch Word, they sustain the etheric fire and the highest degree of Light. The Masters work now to bring a new poetic essence into the worships of men. New symbols and new powers now come into the earth through the new in Ritual.

Joseph, the father of Jesus, who was Noah in a former incarnation, works within the Second Heaven with all religiosity impulses, where all true Ritual supporting religious bodies is sent forth.

Mary works with the Most High Saints and the Mother of the World. Joseph and Mary work together as co-atom Cosmos Souls in the Second Heaven that their son, Jesus, may enter the heart of the one who prays and makes supplication to God. Joseph and Mary are conjoined in one work together in command of the religious impulses—regardless of what a religion is called—as long as Jesus is presented therein as Christ Jesus.

All religious impulses begin in the Second Heaven. All meditative procedures begin in the First Heaven.

The tabernacle in the wilderness correlates to a Sacramental Blueprint in the Second Heaven, where the Sacred Sacraments originate. In the Third Heaven, Jesus, the high priest of Melchisedec and the Sovereign of the earth, reigns over these Sacraments.

When man thinks of ritual as duty, ritual becomes as stone, dead to life in Spirit and in Christ. Man-made ritual carries not the Light, nor does this ritual build onto the high. Through formal ritual, the worship of man becomes rigid, and he turns to separateness.

Ritual without love brings a cruel weight upon the soul. When ritual becomes duty, it can be sustained for only one life's expression. On the coming of the next life, the karma of the old, as duty, oppresses—and man bends away from that which is ritual. So it is today, many in the world observe not ritual; for, in their olden karma, duty oppressed them as worship.

Men will come to know the voice of Logos. In their minds and hearts, they will create the rituals or true formulas after the kinship in Christ. The downpouring of Light and Love comes from the Spiritual Worlds, and those ready to recognize the exact formula and pattern will receive the worship of the new as the New Age begins.

There appeared to men in Earth in the late 1860's a

method of receiving which was of the psychic. Men bringing forth such things brought an injury to the etheric body and a linking with the astral world. On this, men brought a partial in timing, in thinking, and affected the physical health as well as the mental.

That received by pure Seers, by great Initiates, and by all who have brought Light in the Earth has been received in the great Scrolls of Light. And as of writing, it reflects itself upon the brain of man until it etches and burns itself into his consciousness as his own imaging, until finally the idea is sealed in and he begins his work in Earth as expressing.

The Masters in Light and the Great Beings work from one century to another to accumulate and send forth an idea with each age which changes the evolution within the age. This idea is imprinted on thousands of minds and sifted and selected until it becomes as a fire in the minds of men and begins to burn away the ignorant and dark. So are the books of the Archive. So are they enscrolled in the records onto man's receiving and man's aiding onto others who walk in darkness in the material world.

All of the great spiritual dwelling places in the Spheres of Light have been built through the Ritual in the thought world and emotional devotion of men. The great sacred Fountains' prayers and devotion through Ritual bring forth the energy as a sustaining grace on men. With passion, men have painted great mysteries and pageants in etheric light and spiritual design; each painting portrays a Ritual or a key or a secret pertaining to the true things veiled from men of material vision. Each book as it begins to come from the Scrolls of Light carries with it a tone, a sound, a Ritual, by which the souls of men are opened wider in the revealing.

Men serving the Altars of God and preaching in the ministry will no longer speak words of repetition, but will turn to their true soul's Light and speak by inspiration and Love, and will turn to grace and power in the Light of God. They will cease repeating the thoughts and theories of other men and will bring the true Essence from the Spiritual Worlds.

Ritual is not just a set of numbers, a set of symbols, or a combination of words, but is a steady turning to the One, to God. The blending of the One, to include the All within the All, is the true Ritual.

He who lives in Ritual goes ever back to the One, to the All. He carries his vision onto the Great Eternals, onto the endless as to man, and onto the Infinite as to God. And in his turning forth, he finds his freedom in Light. He finds the opening vista between himself and God.

> *Men fear holiness, but men shall come in simplicity to know there is naught to fear in that of God. Only the dark bringeth fear. And the Good bringeth naught of the dark.*
>
> *There is no time or age in the judgment of man's actions in earth. There is only that which he recordeth or writeth within his heart from his actions of love and trust of one another. That alone goeth out into the earth, seeth above the mountain, mingleth in the tree, followeth the quiet stream, heareth the call of the bird, seeth the golden Light of the Christ at sunset, sitteth on the rock and maketh a poem, or walketh forth to greet a friend, or sendeth forth his telepathy as a word-tone of love in which naught can separate or tear asunder.*

ARCHAIC DREAMING AND THE SECOND HEAVEN

> *The Second Heaven contains the Greater Cloisters of Heaven where one meets the Most High Saints and the Great Masters, and receives instruction in the Hall of Wisdom or Shamballa.*

Adepts are initiated in dreams through the seven planetary Logoi or the Archangels over seven planets. These prepare the adept to make union with the Third Heaven.

Initiation into the Second Heaven introduces many unknown or unfamiliar facets into one's dream experience. Through Second-Heaven initiation, dream life moves beyond the sentient or Freudian experience into a majestic contemplative state of dreaming. In the contemplative dream, one encounters the age-old archaic reflective portals, wherein he touches the virility of valiant deeds still living within the soul-memory of mankind. The archaic soul virilities contain urgent, potent and dynamic intelligences. These seek to unite with one living in the world who has a specific purpose or gift to give to mankind.

The archaic virilities are powerfully projected upon the thought screen of the one dreaming. Any person exposed to the archaic-virility dreams receives into himself the power to reproduce the archaic virilities into a vision and a hope for the perfection of man.

An initiate's progressive response to the archaic dream leads to an enlargement of human nature, soul nature and supernatural powers.

Archaic dreams are remembered and recognized by (1) a knowing and confirming of angelic response; (2) identifying oneself with the law of equation; (3) rationalizing one's thoughts through cognition; (4) knowing oneself to be a centered self within a system of spiritualized manifestation.

One's Recording Angel times to him this state of archaic dreaming. The Recording Angel holds aside the veil of the archaic verities when one has insulated the brain sheaths with the mobile ether, and also when he has achieved coordination between the free flow of the will, the imagination and the memory. Through such initiations, dream receptivity is increased. Dream prophecies, accompanied by true interpretation of dreams, are activated. Such interpretation is the only ethical method of dream interpretation. All other levels of dream interpretation are speculative, probing and intrusive.

Interpretation of dreams out of timing can produce a symbology flood tide of catastrophic proportions. Through ignorant probing into dream meanings, one may falsify mesmerically and produce painful reactions to dreaming. This will cause one to fear to receive his dreams, or even produce insomnia or sleeplessness.

When one has been fully initiated through the archaic dream portals, he may ethically penetrate the brain sheaths and the soul sheaths protecting dream imagery or symbols. If one has knowledge of the motivating compulsions of man, which can only be opened through archaic knowledge, he can perceive the root compulsions directing all men. Should he seek to interpret dreams before being initiated in this manner—having only his own judgment and reasoning faculties—he will judge according to his own nature, and thus intrude upon the eternality revealed in the dream. Such dream interpretation can give no lasting help to the one dreaming.

It is God's intent that each man interpret his own dreams through a perceptive unification with his own conscience. In this time or period, unfortunately, many are separated from their conscience due to negative action and works. Through the desire to unite one's self with Truth and with the Real, the Guardian Angel and the Recording Angel will work to initiate one to a perception into the third and little-known aspect of one's nature—the soul and the dream.

O blessed Angel of the night, lead me to the peaceful place, for I would shut away the clamors and the din of a fevered day.

MARY

And the angel came in unto her (Mary), and said, Hail, thou that art highly favored, the Lord is with thee: blessed art thou among women.
—St. Luke 1:28

Behold Mary, whose devotion in Heaven to God intercedes for the unknowing unborn and for the unknowing born. She who gave birth to the highest Being of the earth became the holy example for all life coming to birth. She who grieved the most grievously comforts the grieving. Her solace is unceasing; her helps perpetual. Call on her; she is instant, for all calling penetrates her hearing as a sound through which the voice of her Son, Jesus, is heard.

In the Venus reflected Sphere of Light in the Second-Heaven, Mary reigns with supreme love, mediating with her Son. Her forgiving garment or mantle spreads over the grievous wounds of man—healing, interceding. Here, she lifts, heals, anoints, cheers, loves. Here, she reigns over the Most High Saints—assuring, comforting. To know the Mary Being is to receive the supernatural manifestations. All who turn to Mary receive supernatural helps. All men, women and children of the earth are the sons and daughters of Mary. Mary, the Mother of Jesus, belonging not to any particular religion, hears the extreme call and sends forth her intercessors to do the work of Him so nigh in soul, in heart, in mind.

The right hand of Mary is the hand of Mercy and Law. Her left hand is the hand of Welcome and Love. Justice dwells under her right hand; and Wisdom, under her left hand. Mary is the soul of all women embodied through life after life. She is the Life-vigil, as her son is the Light-vigil. She is the treasure of the milk of instruction, preparing the unteachable for the meat of the Word. Her vigil is constant. Her name is star-behold.

One should avoid making of Mary, the Mother of Jesus, an idol. It is not the soul-desire of Mary that men use her as an object of worship.

Mary works with the Most High Saints and the Mother of the World. Joseph and Mary work together as co-atom Cosmos Souls in the Second Heaven that their Son, Jesus, may enter the heart of the one who prays and makes supplication to God. Joseph and Mary are conjoined in one work together in command of the religious impulses—regardless of what a religion is called—as long as Jesus is presented therein as Christ Jesus.

Mary works to give one Chastity so that he can receive the Immaculate Conception in the heart.

Mary, the Mother of our Lord, as intercessor and mediator for her Son, belongs not to any special church. She belongs to all men who, when burdened, would call upon her to reassure them of their mortal and immortal worth.

Mary, the most chaste of women in the earth—chosen of the Father to give birth to our Lord—is present with men as her Son, Jesus, is present.

Mary's divine intuition takes over the household of our emotions. She dwells nigh to the heart threshold of the wounds of love. She heals the bruises of love misused. She mends the broken places offended by misplaced love. She trims the vines of love entanglements. She washes away the darkened passion-soot accumulated on the flame wick of love.

Mary needs not beseeching. She is already so encased and involved in our love impulses we need only to think with a desire to be disciplined, directed, and led.

Through love, Mary will come with her grace fruits of peace, showing the face of love's requitement.

O thou Mary, how well thou didst see the need of thy awkward penitents. We have walked in the garden of thy love, crushing and bruising the petals of thy heart. Abide with us, thou Mary constant.

Thou sayest to us with thy penetrating mind, "O bliss walkers, come down; walk in the world where grief and gravity travail. Enter into the sacred earth where men struggle, strain, and strive. Remember the carpenter's bench where my Son labored; remember Joseph's instruction to his Son, the builder:

"Come, make thy households my hearth; glorify and simplify the homely tasks of thy hands. Is any genius greater than that of making others happy in comfort?

"Yield up thy frictions; douse thyselves in the purest of oils healing the itching of thy wayward senses."

We turn to thee, Mary; help thou the world. Our souls are pitiful. Our minds weighted by the rocks of stupidity are filled with useless care.

Help us, O Mary, to give a youthful vigor of equal conscience for the young, for the unknowing. Let us know, that we may give to the unknowing.

THE MOST HIGH SAINTS

Let the saints be joyful in glory.
Psalm 149:5

All sacred objects of worship—crosses, cloths, chalices, sacred books, sculptures, water, wine, bread, paintings, jewels, icons, articles of the Altar, baptismal fonts and kneeling rugs—are filled with saintly emanation of the Saints of the Most High. The Sacred-Emanation Saints work with the creative tone sounding within all sacred objects. The Cherubim Angels of the Shekinah work with these saintly hosts, keeping the tones of the objects in alignment with the tones of the Musics of Heaven. When these Saints are present in worship places of dedication and purity, worship is communion. All sacred objects accepted by the Most High Saints sound their correlating harmonies. The power of the Altar is magnified and glorified, becoming a body for the Holy Ghost.

Sacred-Emanation Saints also emanate a supernatural light and life into homes, temples, springs, trees, flowers. Any object or person having holy resonance can be used by these Saints to increase exalted witnessing for God. The sleeping room of a pure, holy person is filled with this saintly emanation. From this one can receive healing. Clothing, dear talismans or objects possessed by a holy person are saturated with these sacred emanations. Saint Paul emanated this holy emanation or magnetism; his very clothing and handkerchiefs healed instantly when touched.

Give me a sign, O Saint; a sign of thy presence;
a sign known between your soul and my soul.

The Most High Saints are the inhabitants of the seventh or apex Cloister. This Cloister is called the *Cloister of the Holy Precipitation.* To approach and be healed by these Saints, one must have released all desire to own, to have, to get, to keep; all desire for authority, for power; all manipulation of objects or things or persons. These Most High Saints will not respond to the devious heart, to the indecisive mind, to the covetous or to any one having desire to exalt himself rather than God.

To ask mediation from this Cloister when one is unworthy will result in magnification of all faults, blots, flaws, stains, sins. The answer to one's cry for intercession will send corrections, dissolutions, disciplines, cleansings, labors. When grace is present and one seeks to receive help from the Cloister of Holy Precipitation, he will receive bread where there has been no bread; he will touch and eat bread manifested through precipitation. He will see objects manifested where objects have not been. And he will be in a state of Supernatural Grace.

Where the Most High Saints dwell, there is hope, for they do mighty things when I call.
He has set the Most High Saints on a throne

nigh unto His glory; their place is a place of the halleluiahs.

The Seraphim and Worship

In the year that king Uzziah died I saw also the Lord sitting upon a throne, high and lifted up, and his train filled the temple. Above it stood the seraphims: each one had six wings; with twain he covered his feet, and with twain he did fly. And one cried unto another, and said, Holy, holy, holy, is the Lord of hosts: the whole earth is full of his glory.

Isaiah 6:1–3

The Lords of Mind working with the higher mentality of men are called *Seraphim*. To be initiated under Seraphim is to prepare for illumination under Christ.

The Seraphim-Angel overlords, perceiving the initiatory trials of the higher mental initiate, come forth to aid and to counsel one who has mastered the lower phases of sentient thought.

There are eras or periods of enlightenment for advanced egos in the world. During these times, the Seraphim Lords of the mind work with certain egos, that they may give to mankind advanced ideas. These ideas stem directly from the Greater Archetypes and are the salvation for mankind. Only men having selfless desire to instrumentalize themselves as servers for God may sustain this very special apprenticeship.

The Great Seraphim are Angels of Erasement. And the Great Cherubim are Angels of Manifestation through form. Before one can come under the Archangels and receive directly from the Greater Archetypes, he must go through the gates of the Cherubim and, finally, encounter the Seraphim. The Great Seraphim take one into the Transcendental.

The Seraphim have six wings which they use to erase the

solidified karma on the first six planes of the astral world. This erasement enables a person to reach the seventh plane and thus enter into the Second Heaven.

The Seraphim are the closest Angels to the Holy Spirit. They carry the Holy Fire, which is the cleansing Flame of God. They guard His Holy Precincts of Altars, of Meditation-Counselings, of Mediation, of Right Affinities. They also carry the fire in the Holy Writ, whereby Scripture lives and survives. The six wings of Seraphim pertain to six ways of writing, six ways of teaching.

The Seraphim keep alive the Esse or the Holy Spirit within the Holy Writ. Through their Angel-tides, they move the Initiate in and out. They are the Guardians of the Threshold of the Altars where the Holy Spirit is manifested.

The Seraphim are the Treasure-Guardians of Soul-Gifts, of Soul and Holy Powers. To be in the Seraphim encirclement or holy encasement is to know the Holy of Holies, the true Shekinah, the true Upper Chamber, the true Archetypal Throne.

The Seraphim are in connection with the Great Musics and the Archetones within the Archetypes. He who has command over his own nature has natural grace, where the maintenance of his body is cared for in both the Archetypal and Duality Kingdoms. One who has knowledge of the cyclic interflows is in command of certain life-forces which are in his own natural-grace nature.

The senses of man are his instruments to interpret his biological origins and purpose. The soul of man is his instrument to intuit, to experience, and to respond to his eternal being. Worship of God is necessary to the soul and to the self. Worship or communion with God returns man to his spiritual origins. Worships based upon primitive impulses stem from biological urges and result in union with the elemental, magical forces. Such worships are fanatical, excitable.

Until the time of Jesus, the elemental forces influenced

the priesthood. The Seraphim Angelic Kingdom was experienced by the greater prophets. Mass worship was sustained by elemental magics, keeping men bound to idolatry, blood sacrifice, sex orgies and bestial practices.

When Jesus came to earth, the Hebrews, though recognizing the One God, still used these teraphim elemental practices in the sacrifice of animals. After His death, Jesus opened the Second Heaven to man and the Seraphim Angels took command of the worship rhythms in the world.

When one becomes aware of his soul and the need to worship, by trials, tests and challenges he is exposed to the inner dynamics of the spiritual and worshipful life. To live in the spiritual life one must master the biological side of life, the soul life and the life of the self. The twenty-one lower astral initiations teach him to face the life of the senses with emotional maturity. The twenty-eight soul-initiations of the First Heaven teach him to center himself in the use of the gifts of his soul. The seven great initiations through the Second Heaven, or planetary reflected Spheres of Light, teach him to use his mind, his thoughts, and his power of creative imaging with ethic.

The Seraphim Angels initiate man into techniques of light, whereby he comes in contact with the Archetypal ideas as given by the Christ Mind.

The Seraphim Angels lead the mind of man into the six dimensions or the door to the Third Heaven, where he is then initiated by the Archetones under command of the Holy Ghost.

Godly men are as wise as the serpent and as gentle as the dove; wise as the Seraphim Angels who are the Holy Nagas of the Great Purusha Unconscious. When the Nagas are gathered together, the mysteries are made plain; the Third Heaven spills itself through the Archetypal Light into the souls and minds of men.

The Seraphim skim through the atomic energies of Space,

free, related to all dimensions. Angels stand upright; Seraphim glide. Mind currents coming to one through the Seraphim are beyond the ecliptical pressures of gravity. The Archetypes are stepped-down to advanced Initiates in these lightning-like, penetrating, gliding movements into the mind. One receives via Buddhi-levels of his own consciousness. Encircled around the Archetypal Initiate, the Seraphim make their home when a special Dharma for New Eras is born.

In the Apostle's Light, all disciples count time from the Christ. All other time and times prepared for Him. In the time of the now, the Increase of God is of the Christ.

Seraphim: "I will pour down upon you the hail of the holy writ of learning, that in the latter days you may be sanctified in the Next Day. In the Archetypal Light, I will inscribe into you the Plan and the Image for the Next Days. I will enscroll into your inward parts your next testimony as a prophecy and as a column of enlightenment."

Thus saith the Seraphim who gives the holy writ for the Ritual of Worship and the Ethic of Works.

"I, the Seraphim, am wherever He, Christ Jesus, is. I go before Him, that right resolutions may yield her fruits in time with the Archetypes."

SECOND-HEAVEN INITIATIONS AND THE PLANETS

Since Jesus, Second-Heaven initiation has been open to "whosoever will." Through initiation in the Second Heaven, men develop a hunger to hear the words unspeakable, to see the glory of the centered Kingdom, to enlarge upon their spiritual knowledge and spiritual power.

If one's grace is sufficient, he is ready for initiation into higher degrees of light; and he begins to earn the right to research the reflected Spheres of Light, or Second Heaven, and to study their influence upon the minds and destinies of men.

In the Spheres-of-Light initiations, one learns of his mind and will. In the initiations in the First Heaven, he learns of his soul.

Second-Heaven initiation is the last bridge between the World of God and the world of man. Anything obstructing one's knowledge of God must be cleared away through the greater initiations within the Second Heaven. One cannot enter the Kingdom of the Eternals or Third Heaven until he has learned to walk the "straight way," nor can he become an actual witness to heavenly thrones until he comes with pure hands.

Jesus as Mediator in Heaven spreads His garment over all Heavens; He stills the tumults and He commands the diversities. No one can storm Heaven or gain access to Heaven save he be initiated in the necessary initiatory labors to gain Heaven. Many in the world thinking themselves to actually see Heaven are seeing but reflected aspects of Heaven. To behold the Kingdom of God, one must have the single vision in Christ.

Second-Heaven initiations manifest in one's life as a wonderful sense of freedom, of joy, of accomplishment. All great ideas which aid mankind are first experienced in the Second Heaven. All greater methods of healing the human body originate in the Spheres of Light or Second Heaven. All great inventions assisting men in their rise in civilizations are first experienced here by the greater adepts. Adepts, in beholding the Cosmos Realities reflected in the Second Heaven, are inspired to transpose them into tangible form to be utilized by man. The greater musics have their origin in the Second Heaven. The master composers are adept-initiates of the Second Heaven; their music is used to transform men and to free them from their sensual gravity imprisonments.

In the Second Heaven, one perfects the self into a third-dimensional consciousness being. In the Third Heaven, one is initiated into the ethical use of spiritual revelation.

In the Second Heaven, one is initiated into manifestation or Melchisedec powers. Here, he is instructed by the composite telepathy of the Masters or Great Immortals. The Melchisedec powers enable him to become a manifestor in right timing, and to de-manifest negations—with ethic.

The Cosmos Soul of Jesus unites the world, the planets, and all life in Heaven and in earth. The mass totality of an eternity and the enmassed homogeneity of souls are His first concern.

As Jesus was initiated into mastery over the energies within eternal substances, all who would follow Him must achieve the mastery of cosmic or eternity substances. Men of the earth are initiated cosmically. Jesus was initiated cosmosly. Man's work is microcosmic in comparison with the macrocosmic work of Jesus. The miracles of Jesus can be understood in no other way than through one's own direct contact with the powers of transubstantiation. All transubstantiation begins with the power of prescient thought. In Second-Heaven initiation, one increases his power to create with ethic in thought, to envision and image with ethic. In time, the souls having this power will do as Jesus said: they will move mountains, change water into wine, raise the dead, and do other miraculous works impossible to conceive in the present development of the mind of man.

First-Heaven initiations are experienced in periodic, joyous, expanding, confirming states of grace. Second-Heaven initiations are experienced in rhythmic exploration through testings, trials and research.

Second-Heaven initiations relate to knowledge of the constitution of the self and of the mind. The Spheres of Light, or Second Heaven, are vast reflective, floating vortices moving with the orbital action of the earth. These reflective Spheres of Light are condensed planetary receptors. The higher interplay of the planetary action falls into these orbital reflectors. After the First-Heaven initiations are concluded, one begins the Spheres of Light or Second-Heaven initiations.

When some persons first penetrate the Spheres of Light, they think they have penetrated the physical planets accompanying the earth. Such interpretation is incorrect. No person living within this cosmic eternity system travels to other planets in his etheric body. Whenever there is anything to be spiritually discerned about a planet accompanying the earth, it is disclosed through initiation within a reflected Sphere of the planet. This Sphere is called *a Sphere of Light*.

In certain cosmic seasons, such as the equinox and solstice periods, initiates undergo special initiation and instruction within the Spheres of Light. Through the Seraphim Angels, the initiate learns of the earth's creation and of the earth's future manifestation.

Seraphim Angels are the Planetary Logos Angels, or the Lords of Mind. The Spheres-of-Light initiate is overdirected by these mighty angels. Under their instruction, he enters one reflective, planetary, etheric Sphere at a time. Through the Seraphim Second-Heaven initiation, he learns to use rhythmic telepathies in conjunction with the higher energies vitalizing, circulating and stimulating the higher mental and spiritual impulses of the earth. The Seraphim Angels open the cosmic secrets of the mind. The powers of imaging and thought as creation are presented to the initiate in their most creative and positive aspect.

The Planetary Logos Angels are directly under the zodiacal Elohim-Hierarchs of Constellation. The Hierarchs are: Aries, Taurus, Gemini, Cancer, Leo, Virgo, Libra, Scorpio, Sagittarius, Capricorn, Aquarius, Pisces. Jesus is the Mediator between the Hierarch powers. During auspicious or favorable planetary aspects, the Planetary Logos or Seraphim Angels enable the initiate to travel etherically to the reflected Spheres of the planets accompanying the earth. This reflected Sphere or Second-Heaven action is sometimes called *Shamballa,* and in the Scriptures is called *Salem* or *the Holy City.* Melchisedec and the Ancient of Days are the initiators in this environment. Here, Maha Chohan and the Great Immortals also work.

The Second Heaven, or Spheres of Light, is located in a celestial, orbital atmosphere outside of the earth. These Spheres are correlating reflective Spheres of seven planets accompanying the earth. When one is initiated in the Spheres of Light, he is initiated through the planetary reflective Spheres of Mars, Mercury, Venus, Jupiter, Uranus, Neptune and Pluto.

The Second-Heaven initiation is experienced by the initiate first through the reflected Sphere of the planet Mars; second, through the reflected Sphere of Mercury; third, through the reflected Sphere of Venus; fourth, the reflected Sphere of Jupiter; fifth, Uranus; sixth, Neptune; seventh, Pluto.

The Spheres-of-Light initiations are not fulfilled in one life. Many lives over long periods of purification through the interior Saturn initiations and the lower astral purgings are required to attain the Seraphim crown.

Mastery over the lower planetary energies falling upon the magnetic belt around the earth—and the mastery of the kinetic, frictional, electric, electronic, electromagnetic forces energizing all earth forms—is accomplished through the will initiations or the power to raise the will above sentient-will action. When one has incorporated the will to love life, the will to love persons, the will to love God, he enters the Second Heaven to study the reflected Sphere of Mars. Here, he learns of the Hierarchy overdwelling the planet Mars. He sees the planet Mars as the energizer for the blood of man. He learns that were it not for the ebb and flow tides between the Earth and Mars, man would be either a horizontal creature or a floating, gravity-free etheric form. In the Mars reflected Sphere, he learns that the spinal canal is the will lightning-rod of the body, assuring an upward reach of his mind and body.

Through Second-Heaven research, one sees that the negative Luciferic influences affecting man are made possible through the lower electrical and magnetic influences of the planet Mars. He also sees that all that man may do courageously is made possible through the higher influences of the planet Mars.

During Second-Heaven initiation into the reflected Sphere of Mars, the initiate must attain and sustain a mentality of peace. He sees and observes that science without love is stones minus bread. He sees that men must cease to disagree as to the basic realities of the Universe, and that the greatest error is to insulate or isolate knowledge. He sees the raw unharnessed elements to be awaiting a science based upon spiritual reality. To become a scientist of spiritual reality, one must fix his mind upon the uncontroversial eternal design.

The Seraphim revealment teaches the initiate that the physical, bacterial, organic, sentient, or mental stages of evolvement require struggle and effort. The initiate is placed in the center of this struggle, that he may confirm what he learns in the Second Heaven. He sees that matter is congealed light and that obstruction in matter produces friction, fire. He learns that the destroying Omega-tones keep matter in a mobile state of fire. He sees that all sentience contains irresistible fire animating mineral, plant, animal and man.

The eye of the Mars initiate looks into the fires of the earth, and his vision beholds the Light of God permeating all things. He understands the sins of omission and inertia. He knows that what God cannot use He consumes in the flame of purification.

To study the planet Mars in the reflected Mars Sphere, one must become a fire-initiate or will-initiate baptized by the fire of the Holy Ghost. To be cleansed in this fire qualifies the initiate to come under the instruction of the Great Immortals who, as fire initiates, followed the star of Bethlehem to the cradle of Jesus. Through the Holy Ghost, one learns of the power of the blood in man. Particularly does he observe and seal into his mind the power of the blood of Jesus shed for man; the power of the Eucharist Sacrament or wine transubstantiated from the blood of Jesus to the blood of man. The Mars initiation teaches him to see the blood of Jesus flowing through the veins of all men, thereby assuring the human spirit of the communion between the bodies, hearts and souls of men.

All that man may know of his mind as applicable to his living in the world—all that he may do through thought—is observed and utilized through the initiation in the reflected Sphere of Mercury. All thought, manifested and unmanifested, concerning man's life in the world of the living and the dead is reflected in the Sphere of Mercury. And finally, all that is spoken or yet to be spoken is confronted here. In the reflected Sphere of Mercury, the initiate learns that thoughts unspoken belong not to the speaker. Thoughts, when spoken with creative knowing or animation, become objectified forms having life. Such thoughts are the progeny of the speaker who has quickened them. Jesus knew this power; and He passed this power on to His Apostles, that man might receive the authentic embodiment of His abiding word.

Men crave love, seek love, die for love, knowing not the simple, sweet, uncomplicated way of love. This is the love Jesus knew for men. *The will to love is the way of love.* He who has not the will to love is dead to life. The Commandment of Love is a will testimony by the Master of Love—Jesus.

In the Venus reflected Sphere, one is initiated into the

rhythms of Seraphim. The rhythms of Seraphim move as tides of love engulfing the earth and its peoples. In the Venus reflected Sphere of the Second Heaven, one comes face to face with the Most High Saints; with Mary, the Mother of Jesus; with Joseph. In the Venus reflected Sphere, one learns of the love of "greater love hath no man than this, that a man lay down his life for his friends." (St. John 15:13) The Seraphim tides teach men that love without the self is the lamp without oil, the bread without leaven, the hearth without the flame, the priest without an altar.

Before Joseph became the father of Jesus, he lived as Hiram Abiff in Solomon's time; he assisted Solomon in the temple initiations. As a master adept of Tyre, he received the record of Atlantean sacraments in the world. In the present time, he is a Great Immortal working with the Sacramental Cloisters in the Second Heaven. He helps men of the earth to keep alive the supernatural and the pure in their worships. The Sacramental Cloisters overdwell the differentiated expressions of religious worship in the earth, assuring their evangelistic virility and their ethic.

Those initiated through Joseph and Melchisedec, the king of Salem, become priests of Melchisedec. Their devotion and dedication sustain the fire or flame on the altar place, that men may lose not their faith in God.

In the Second Heaven, Mary and Joseph overdwell the Most High Saints. Those who look on Mary or Joseph in Heaven as Deity, or on any Being as their Deity, save God, are blind to the intricate sensitiveness of Heaven. There is one Deity—He is God, the Supreme and Eternal Spirit.

In the Second Heaven, one comes to know Mediation as it is in Heaven and in earth. All in earth and in Heaven are gradated mediators for God. To be aware of the network of cognitive, intelligible energies is to know the science of Infinity.

In the parable of the Wedding Feast (St. Matthew 22:

1-14), Jesus tells of the necessity to be qualified or to be prepared for participation in the Kingdom. To earn the wedding garment, one must be initiated through the Spheres of Light or Second Heaven. The wedding garment is the true garment or spiritual body worn by the soul. The wedding garment is a seamless garment, an eternal undying garment, not made with hands, but made with the skills of the soul.

To open the door to the mystical Presence of God, to go in and out, to become a householder in the many mansions or Spheres of the Second Heaven, one must be willing to live in the world while preparing to become a citizen of Heaven. A spiritual largesse must enter the mind, heart and soul. One must learn of magnanimity. He must be free of claim or possessions; he must give and forgive; and he must walk on the path of the white, sharp stones—or sacrifice. Jupiter-Sphere initiates have trod this path and are ready to learn of expansion; they are ready to extend their powers of manifestation and de-manifestation.

The true charity and soul-concern for men in the world; a charity beyond token objects of pity; a charity based upon a charismatic intuition of the true need in the giving of the bread—this is the mystery revealed in the Jupiter Sphere of the Second Heaven.

The reflected Spheres of Saturn, the Moon and the Sun, having distinct initiations of their own, are not part of the Spheres-of-Light initiations. The Moon's reflected Sphere is activated in the lower and higher astral world. When a person is initiated in the Moon's reflected Sphere, the astral world is clarified to him, and he masters the currents of the astral world.

Saturn's reflected Sphere is located and activated in nine of the thirteen etheric layers of the earth. When one is initiated within the reflected Sphere of Saturn, he has (1) an initiatory experience within nine etheric layers of the earth;

(2) he becomes acquainted with the nine orifice atoms of his physical body; (3) he learns of the chemicals, minerals, liquids and solids of the earth; (4) he researches his skeletal structure; (5) he is initiated into the memory of the generations of mankind; (6) he is initiated into form; (7) he is initiated into the power of levitation.

The Sun's reflected Sphere is activated in the remaining four etheric layers of the earth and centers itself in the core of the earth. When one is initiated within the Sun's reflected Sphere, he is initiated through the trials of discipleship correlating to the Realm of Light and the Archetypal Kingdom.

In the initiations within the Moon's reflected Sphere, one learns of his emotions. In the initiations within Saturn's reflected Sphere, he learns of his physical body. In the Spheres-of-Light initiations, he learns of his mind and will. In the initiations in the First Heaven, he learns of his soul. In the initiations within the Sun's reflected Sphere, he learns of his spirit.

The Moon's-reflected-Sphere Angels are Propagation Angels, the Race Angels, the Lunar Angels and the lesser Cherubim Angels. These Angels work with the Race Lord Jehovah through the phases of the Moon.

The Lunar Angels are the Angels over the human blood. They work with the genes and the chromosomes. They also work with races, tribes, foods, grains and the waters.

The Moon's reflected Sphere is part mirage and part luminosity. The mirage mirror of the Moon's reflected Sphere reflects the negative race-memory in the lower astral world. The luminosity Sphere of the Moon reflects the combined soul-grace of humanity. Jehovah and his Angels are in command of this luminous reflected Sphere of the Moon.

The greater Cherubim and the Terrestrial Angels work with Saturn's reflected Sphere. The Seraphim Angels work with the Archetypal Kingdom or the Third Heaven.

Saturn's-reflected-Sphere Angels are the Greater Recording Angels and the Angels of Judgment. These Angels determine how much man may remember of the eternals. They also work with the world conscience.

The reflected Sphere of Saturn is an electromagnetic *Cosmic Sphere*. Saturn has more influence on the various planets of this eternity system than any other planet. Saturn is a lead-laden ballast and a monitor for the other planets and for man. Saturn assures the equation for man and also determines the pace and the movement of the other planets accompanying the earth.

The Father works with man in all Saturn initiations. Through Saturn initiations man comes face to face with the Father. He learns to adore the Father and feels compelled to worship Him.

In the beginning of the creation of this eternity system, Saturn was the first planet ejected from the earth. The Saturn reflected Sphere in the first nine etheric layers of the earth is the door to initiation. All who enter into the Spheres-of-Light initiations must first undergo the Saturn initiation. To be qualified for higher initiations, one comes under the influence of the somber Saturn tones which play upon the higher etheric body, the memory portal of the mind, the heart and the procreative organs.

On entering the Saturn reflected Sphere within the earth, one is initiated through the somber, mathematical, coordinated sounds of inner-world music. Through magnified sensitivity in dreams and sleep, one hears the Saturn tones and begins to learn something of coordination, enabling him to gauge the true dimensions of bodies and forms; he learns to differentiate between objects created and uncreated. Thus, his discrimination powers are refined and quickened as to the law of relativity and perspective. Through the Saturn initiations, the undeveloped mystic overcomes psychical scatteredness.

Through the Saturn tones, the mineral life of the physical world is recorded upon the lesser and higher etheric bodies of men. In the Saturn sound-initiation, one learns to hear the breathing soul-life in the seed, in the sacred atoms of the plants and trees. He learns to move through indestructible confinements. He moves beyond the barrier of physical sound into the harmony of order, discipline. He overcomes the feeling that he is the only one. He learns to hear his own soul's tone, and knows that it is but a microscopic unit with countless billions of soul-tones in this earth system and in the greater Universe.

On completing the initiations in the Saturn cosmic Sphere; in the Moon mirage and luminosity Sphere; in the Mars, Mercury, Venus, and Jupiter Spheres—one is then initiated into the Uranus, Neptune and Pluto Spheres. This occurs in triad action, as these three planetary Spheres are three spiritual planetary avatars stimulating man's will, his mind potential, and his higher consciousness.

Jesus, knowing timing as none other of this earth system, established His mediation rhythms to coincide with the accelerated works of Uranus, Neptune and Pluto. The Son of Man saw the rising tide of man's ripening consciousness and He foresaw the mental power of man to come. One phase of His work was to open the Second-Heaven initiations to the soul of man. His words, deeds and promises were based upon the knowing of the mighty spiritual and cosmic tides to be eventually utilized by the minds of men.

In the Uranus, Neptune and Pluto Spheres of initiation, the initiate learns that man is indeed a god or hierarch in the making. *"Jesus answered them, Is it not written in your law, I said, Ye are gods?" (St. John 10:34)* The initiate beholds the coming mental capacity of man as being equal to the mind of Jesus. He sees the will of man as an undeveloped and neglected instrument now being used for self or sense gratification. In the Neptunian initiation, he learns how to use the spiritual, etheric, spinal will-fire with ethic; he learns of the transforming power of the will. Through Uranus initiation, he takes command over energies; he is shown how Jesus changed water into wine, how He multiplied the loaves and the fishes. Through Pluto initiation, he looks into the unborn or unmanifested impulses surging and seething beneath the portals of man's subconscious mind; and he learns of man's ultimate victory over death. Through a gradual process the initiate beholds the man of the future, when science will acknowledge the influence of planetary energies contributing to the constitution of man on earth.

"And ye shall know the truth, and the truth shall make you free." (St. John 8:32) Jesus came to unite all things of the earth with Heaven. He is the revealer of worlds known and unknown. In man there is a mind similar to the Jesus mind. In time, all that the "Father hath" shall belong to man. The initiation processes through the Spheres of Light or Second Heaven are the openers of the door to God's perfect Kingdom and a preparation for producing a mind under Christ.

> *The Guardian Angels protect the records of the souls of men. The Recording Angels protect the record of the works of men. Only one who has been initiated into the Spheres of Light, and has overcome the 8th Sphere, may open these records and see revealed both the record of the good and the record of the dark.*
>
> *World Grace is Grace earned by humanity through good works in this eternity. World Grace is watched over by the greater Recording Angels known as the Angels of Judgment. This Grace enables mankind to experience its Saviours. If one has alignment with World Grace or the works of good of mankind, he is an adept who has commanded the gravity trials, and has made contact with the Planetary Guardian Angels of the Spheres of Light or Second Heaven. He thus works with the fulcrum energies at midnight and also with the solstitial and equinoctial energies during the four seasons of the year.*

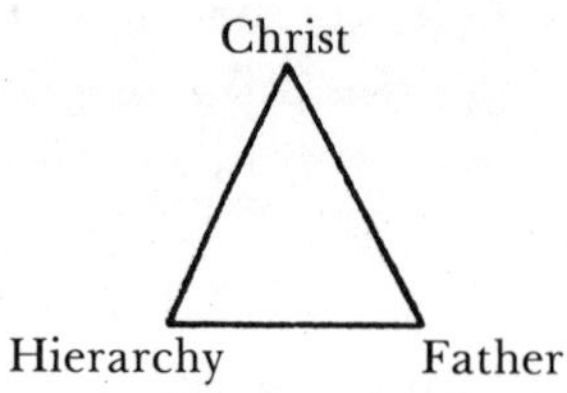

ARCHETYPAL KINGDOM (Archetone or the Great Silence)

Angels Celestial Beings

The *Archetypal Kingdom* is a Cosmic Sea of Virginal *Archetypes* or Blueprints of God waiting to enter the minds of men.

The Christ, the Father and Hierarchy, working with God, are in command of the Kingdom of God for this Earth.

REALM OF LIGHT (Light)

Angels and Archangels

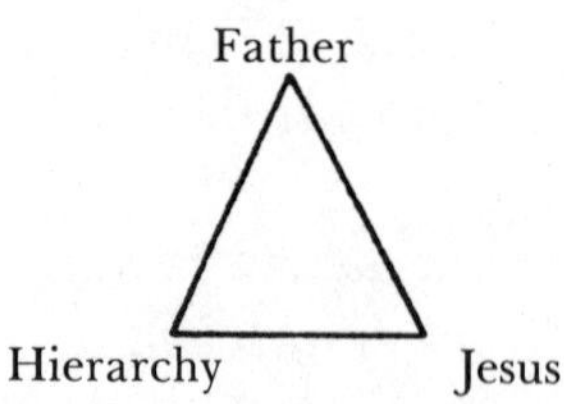

The Realm of Light is the homeplace of the Lord Jesus and His twelve disciples who aid Him to translate the Tone and Ideas of the *Archetypal Kingdom* into Light; to translate the Prototypal Rays of Hierarchy so that man on Earth may become like Jesus; and also to translate the higher energies of the planets, that man may establish the perfect Ethic within himself.

THIRD HEAVEN: KINGDOM OF GOD

15

THE THIRD HEAVEN

I knew a man in Christ above fourteen years ago, (whether in the body, I cannot tell; or whether out of the body, I cannot tell: God knoweth;) such an one caught up to the third heaven. And I knew such a man, (whether in the body, or out of the body, I cannot tell: God knoweth;) How that he was caught up into paradise, and heard unspeakable words, which it is not lawful for a man to utter.

—2 Corinthians 12:2–4

The Archetypal Kingdom

The Third Heaven is the Kingdom of God: the Realm of Light and the Archetypal Kingdom.

The Archetypal Kingdom is beyond man's limits in Time. The Archetypal Kingdom is revealed through the mastery of Timing whereby man comes into the mastery of Space, and moves beyond the limitations of gravity into energies which are of a finer substance than the gross-electrical timing processes of karma in the Earth.

Paul, an adept of the mysteries of God and an apostle for the Christ, assures the reader of the Bible that the Third Heaven is an actual place, a place where "unspeakable words" are heard. (2 Corinthians 12:2-4) In this Third-Heaven experience, Paul was exposed to the unmanifested Archetypes of Heaven that cannot be heard by the untrained ear, nor spoken to the unknowing mind. All adepts having the spiritual power to enter into the Third Heaven are hearers and seers of the Archetones and the Archetypes of Heaven.

The life of Paul is a testimony of initiatory trials. His searching, young years; his researching of his conscience for the stoning of Stephen; his final meeting with the transcendent, etheric Jesus—destined him to become the thirteenth apostle for Christianity. His suffering, physical and mental; his escape from his persecutors until his work was finished—graphically disclose the way for all adept-initiates in the world. It is the destiny of initiates to come to the world to bring a message; to fulfill a mission; to be received by the knowing; to be persecuted by the unknowing; and to die martyrs, that men may retain the imprint of their holy works upon their consciences.

There are three states of Heaven man must encounter and master: the First, Second and Third Heavens. Heaven is a tangible energy, non-astral, non-physical. Heaven is in a state of containment in its rhythmic and spiritual impulses. Wherever the astral energies exist, Heaven is but one-third in its vitality. Where minds are yet dependent upon the mental body of the masses rather than the mind in Christ, Second Heaven remains in proximity to the soul and the soul-powers.

Jesus, the Sovereign Mediator between the world of man and the World of God, dwells within the Realm of Light. His twelve disciples, who no longer re-embody or return to the earth, also dwell in the Realm of Light.

Jesus, enthroned in the Third Heaven, is the Cosmos, Heavenly Mediator uniting the initiatory labors of the directing Hierarchy over each planetary body accompanying the earth. All initiatory paths lead to Jesus. All sacred instruction stems from His mind and from His mastery over the elements, energies and vitalities of the earth and of the spirit.

The seamless garment of Jesus is the higher etheric body of the earth centered in the Third Heaven. This garment covers the earth as a supernal garment, encompassing the human spirit and the souls of those who hear His voice and know Him, and also others yet to know Him.

The way is long. God has given man eternal days in which to accomplish and to attain. To enter into the Third-Heaven initiations, one must have mastered the worship mysteries through the Second Heaven.

The lesser lights or planetary bodies reflected into the Second Heaven are intermediary cosmic supports for the Third-Heaven thrones or greater Presences of the Third Heaven.

Christ Jesus and the Father, with the Luminaries of Heaven—the Angels, the Apostles, the Celestial Angels, the Archangels, the Hierarchs, the Holy Ghost—send the heavenly light of the Third Heaven into the Second Heaven. The Seraphim Angels and the Holy Presences of the Second Heaven send the Second-Heaven light to the First Heaven. The Saints and all Presences of the First Heaven send the First-Heaven light into the souls and the minds of men.

It is the wonder and the miracle of man that he can receive and experience the varied degrees of Heaven, and he can receive more when he believes more of Heaven. When he uses what Heaven teaches him, he is more than mundane man—he is a mind and a soul aware of God.

To be initiated into spiritual power is one thing; to become an adept of spiritual power is another. An initiate is in a state of observation, experimentation and participation, while an adept is in a state of manifestation. Spiritual power is authorized of Heaven when worthiness, love and ethic are equal. All persons having spiritual power authorized of Heaven are under the Holy Ghost and the Christ.

Persons having only religious creedal power are using magnetic, mesmeric forces. Religious power absent from Third-Heaven authorization produces a psychical stimulus minus spirituality. When sanctification is humanly appointed, the spiritual is absent and the will of man is present.

The adept approaches the Third Heaven knowing it to be holy ground—where the Father watches over the seeds or the first fruits of being; where ideas are of the most high; where no man may enter in save he have the name written in his forehead.

> *"I . . . will draw all men unto me," (St. John 12:32) Jesus is the holy magnet set and centered in the frame of the Third Heaven. "I am the door; by me if any man enter in, he shall be saved, and shall go in and out, and find pasture." (St. John 10:9) Jesus draws all men to his heart kingdom. "I am the way, the truth, and the life." (St. John 14:6)*

"And I will give him the morning star." (Revelation 2:28) Those who overcome the social church and raise it to a soul church acceptable for all nations will receive the "morning star" or Venus-Initiation gifts: harmony, peace, beauty, culture. When this occurs, the morning or the new day will have arisen in the world, conquering the Jezebel astral instruction powers—and sensuality will have ceased in the church. When this church has been established, the old mold of religion will have been "broken to shivers." (Revelation 2:27) And the church will receive its worship im-

pulses directly from the Third Heaven, as did Jesus from the Father: ". . . even as I received of my Father." (Revelation 2:27)

Healing Grace is the Grace of the Love of Jesus, the Lord of Healing Grace, who dwells in the World Medallion. When one has earned the Love of Jesus through spiritual proximity, he has heart-to-heart oneness with the Jesus One, the Lord of Healing Love and Grace. He stands as a mediator in the Christ Light, fulfilling the Christ Mind, and thus lives within the Archetypes expressing the first degrees of Creation. He has access to Melchizedek powers of transubstantiation, or the powers of de-manifestation and manifestation.

THE COMMAND ARCHETYPE

The Third Heaven is the center, the throne, and the equi-point of light where all spiritual light is coagulated and condensed. Here, Christ Jesus reigns. With Him are His twelve disciples and leagues on leagues of mighty Angels. Jesus dwells in eternality, watching over the cosmic ebb and flow of the earth, the planets, the continents, the nations, the peoples, their souls. Here, as Holy Arbitrator, serene in the Father—His mind in Christ, His heart in the human spirit—He knows with omniscient knowing all of the heartbeats, pulse points, cells living and yet to live. These are His circumference, His omniscient terrain. All feeling, all thinking, all willing are as fish caught into His great intermeshing Sea of Mediation.

When man finds the Light of God within himself, the true patterns begin to form. He begins his disciplines and works. Each day a small stylus is placed in his hand and a record is written in the Scrolls of Light. Each day his spiritual stature increases. He makes his mark deeper and deeper into the World of Light. On reaching the First

Heaven, he brings forth the knowledge of Beings greater than himself. He knows then that his work has just begun. He dedicates himself to the Father, and thereafter his work is to establish the Father's Kingdom on earth.

All souls desire to be initiated into the Kingdom of God and aspire to receive the direct Archetypal Revelations. To qualify to know the Kingdom of God, one is initiated in the following: the earth domain; the lower astral world; Saturn's reflected Sphere; the higher astral world or First Heaven; the Spheres of Light or Second Heaven; the Realm of Light and the Archetypal Kingdom (Third Heaven.)

All souls of the earth will hunger and thirst for God—the Living God—until they eventually relate themselves to the Third Heaven. One should think of the Third Heaven or Kingdom of God as a reality. His devotion, aspiration and prayers should be centered upon the thought that in the Kingdom of God are the greater verities and greater realities. *"But seek ye first the kingdom of God, and his righteousness; and all these things shall be added unto you." (St. Matthew 6:33)*

The lower astral world is the undeveloped emotional body of the earth. The higher astral world, or First Heaven, is the grace emotional body of the earth; all is love and soul in the First Heaven. The Second Heaven is the higher mental body of the earth. The Third Heaven is the spiritual body of the earth.

The lower planetary light builds a body for the lower astral world. The combined soul-light of Illumined Beings builds the body for the First Heaven. Hierarchy light from the planets builds a body for the Second Heaven. Archetypal light builds a body for the Third Heaven.

In the Third Heaven there is a Master or Command Archetype controlling all other Archetypes. This Archetype correlates directly to the Cosmos Atom in the Sun. The Christ Spirit is in command of this Master Archetype.

When the Christ Spirit entered Jesus, the Command Archetype was activated in the Third Heaven. The Christ Spirit, using the Command Archetype, works to make all worships pure and to extend the mental atoms of men, that men may become imagers for the Father.

> *The law and the prophets were until John: since that time the kingdom of God is preached, and every man presseth into it.*
>
> —*St. Luke 16:16*

Since Jesus, the beginning of the Self-Genesis Age quickened in man a will under God, whereby man entered into the greater compulsions of the Christ Command Archetypes. Since Jesus, the soul of man is being pressed into Heaven or the Kingdom. Regardless of one's evolvement, the Magnet of the Christ draws all men with a mighty pressure into the Kingdom of God.

> *But all these worketh that one and the selfsame Spirit, dividing to every man severally as he will. For as the body is one, and hath many members, and all the members of that one body, being many, are one body: so also is Christ.*
>
> —*1 Corinthians 12:11,12*

Since Jesus came, Christ affects the life of all men. For Christ is the light of the world, and light is the atmosphere of man's Self-Genesis mentality. Resistance or acceptance is all the same to the "selfsame Spirit" working through Christ in the mind of man. This selfsame Spirit, with varying manifestations, is speaking now. The pressure of the Christ Command Archetype stirs, stimulates and compels all men to look within to find and to live in the Kingdom.

Religions under the Command Archetype work with the Will of God through Christ Spirit. The Father and Holy Ghost sustain these religions. All other Presences work to equate and balance spiritual power within worships.

Religions channeling the Hierarch power of the Elohim

or Host have access to Third-Heaven knowledge. Religions freely giving of the Hierarchy impulse inspire men to become masterbuilders, creative souls while living in the earth.

When the Archetypal Christ Light of the Third Heaven falls into the Second Heaven, man experiences God in a third-dimensional initiation. He sees God as the All, the One. When the Archetypal Christ Light falls into the First Heaven, man sees and receives God through the help of the Angels, Saints and other Holy Beings who help him to behold God as good, as love, as justice. When the Archetypal Christ Light falls into the lower astral world, it is challenged by Satan and his darker angels. When the emotions of man contain more of the astral than of Heaven, man sees not God, but only the mirror of his own self-importance. He sees all things "through a glass, darkly." (1 Corinthians 13:12)

The Angels and the Presences inhabiting the Third Heaven are enthroned in heavenly dominions authorized before the beginning of the world. The mighty Archetypes sealed into the Third Heaven sustain all forming, all shaping, all creating in earth and in man.

In the most exalted reach of the Third Heaven are twelve major Archetypal Blueprints. These contain unceasing, multitudinous designs, images and ideas. Centered within each Archetype is an Archetone, or Eternal Hum. When the Eternal Hum sounds in one of the twelve Archetypes, the unmanifested images are activated and manifested into the world of man. The Being called the Holy Ghost, working with the Will of God, is the mighty quickener of the Eternal Hums within the Archetypes. When all twelve Archetypal Image-Impulses sound as one, the earth will have entered into a finished work, and the soul of man will be prepared for another journey in Cosmos.

> *A true seer is in the Light, and sees only* that Light *which lighteth the soul of every man.*

APOSTLES OF JESUS

Jesus, as the central Sun or thirteenth point in the Constellation, reigns with His chosen in the Third Heaven.

The illuminative state within the Third Heaven is controlled by the Christ and Jesus, the Apostles and all the Beings of Omnipresence who know God as Absolute and Eternal Spirit. To experience the Third Heaven, one must become co-atom to Jesus. Here, the initiate and adept receive the power of Archetypal Light directly through the assistance of the Christ Spirit overdwelling Jesus.

The Lord Jesus dwells in the Third Heaven surrounded by His twelve Disciples, the Most High Saints and the Angels Celestial. The Christ Spirit centers His Light in the Realm of Archetypal Light of the Third Heaven.

Each Apostle contains something of Jesus within him, as Jesus contains the Father. Men on earth correlate to His Apostles and will eventually express themselves through the twelve holy zodiacal attributes. When one enters the world, his zodiacal chart expresses the positive and the negative of his birth sign (Aries, Taurus, etc.); yet, in his natal chart can also be found the emphasis-points of the zodiacal Hierarch and Apostle overshadowing.

The human spirit is divided into twelve prototypal patterns. The thirteenth etheric prototypal pattern correlating to Jesus is centered in the sacred heart atom of all men. To activate and stir the thirteenth prototype is to know Jesus, the Christ, the Father. The more one relates himself to the prototypal reflection of Jesus, the closer he will come under the Apostle instruction from Heaven. When one has finally incorporated the twelve higher Apostolic power-attributes, he will then become "like" Jesus. No man of the earth can accomplish this in one lifetime.

The Apostles of Jesus no longer need to return to earth or

reincarnate in human form. They oversee the postulants, initiates and the adepts of the Christ. Each Apostle is a Prototypal Blueprint for one of the twelve types of men dwelling on earth. Every man on earth correlates to one of the Apostles. Each man of the earth may come under an Apostleship protection and guidance. If he will accept this blessing, he will receive Apostleship powers while living on earth.

Jesus, as the Solar Avatar seated within the Third Heaven, wears the mantle of the Cosmos Son, or the mantle of the Christ. Through the prototypal trials experienced within the heart chakra, man, on being born to the world, moves from one zodiacal sign to another, that he may incorporate all of the traits, qualities and capabilities of each zodiacal, prototypal sign.

HEAVENLY-ORDAINED SAINTS

The true beatification of a Saint is made in Heaven. The Recording Angels overdwelling beatification resound their songs and praises when a Saint is enrolled for beatification. To come under the mantle of a beatified Saint is holy grace, providing a sustained peace and an unalloyed joy.

Previous to the time of Jesus, the anointed and beatified Saints inhabited the First Heaven, even as they inhabit the First Heaven at the present time. After the Resurrection of Jesus, their saintly powers were increased. To commune with the Saints and to come under their protection, one must have earned the grace and also have a heart alignment with the Lord Jesus.

The Heavenly-Ordained Saints are Saints whose Supernatural Cloisters are in proximity to the heart of the Lord Jesus in the Third Heaven.

All true healers of the heart, or spiritual healers who

work with love, are supported by helps from the supernatural precincts of the Saints. The Heavenly-Ordained Saints, being on the right hand of the Lord Jesus, are one of His holy conduits for healing, mediative works.

All Saints work with the Angels and are under the direction of the Lord Jesus. When prayers are sacred, the Anointed Saints receive them as petitions in music. The Heavenly-Ordained saints in the Third Heaven make intercession, and the chorus of the Angels waft upward the prayers and the music to the heart of the Lord of the Heart, Jesus.

ARCHETYPAL KNOWLEDGE AND THE CHAKRAS

> *The true spiritual flow of all reaching to God is caused by the Divine Thrust of the Third Heaven. In true Samadhi, one unites with the Third Heaven.*

The lesser yoga teachings confine their research and practices to seven chakras. The great illumined souls of the Royal Yoga line point the way to union with five active chakras *beyond* the seven initiatory chakras. These chakras are involved in the practice of Samadhi meditation; and only the rarest of souls have pierced the veil of these chakras functioning in the higher degrees of Illumination. Those who have made contact with these superconscious chakras through meditation are more often unable to articulate or put into words what has been experienced, as in the case of Saint Paul who, during Samadhi, was caught up into the Third Heaven and heard "unspeakable words."

The five higher chakra Soul-portals, or superconscious chakras, are opened to those who are illumined under Absolute or Unconditioned God. Untapped Archetypal knowledge is experienced, and men of the earth are given revelations to be obtained only in the opening of the five higher chakras. These five ultra-energy superconscious chakras are now to be explored and mastered by the illumined soul of the West.

The Christ Spirit, speaking into the minds of those called to teach and to reveal, gives the greater Archetypal knowledge to those having unblemished vehicles. The Christ Spirit is beginning to enter into the divine wave lengths of light and illumination of those who are willing and ready to serve, to heal, to teach and to reveal. Samadhi under Christ will enable pure and true initiates to articulate and activate what is experienced in the true Atma Light of Unconditioned or Absolute God.

THE FATHER'S WILL

The Third Heaven or Archetypal Kingdom overdwells the greater sacraments in the Second Heaven and the akasic records or world-memory in the First Heaven. The *Greater Archetypes*, sounding their creation and manifestation hums, are silent to the ears of those who are not ready to see or hear what the Christ Spirit has said through Jesus. This silence is called the *Greater Silence*.

The Greater Silence is a Cosmos, Heavenly Sea containing Blueprints of all things unmanifested. The Greater Silence is a supernal and holy opposite of the three turbulent abysses. The substance of Heaven being light and vibrancy (rather than vibration), in the Third Heaven all sound, as men know it, is manifested as a great and mighty hush. When one has been found worthy for Third-Heaven initiation, he must be initiated into this spiritual hush or the Greater Silence of Heaven. Then only will he be given a language and a tongue with which to interpret and reveal what he experiences and sees in the Greater Silence.

Through Third-Heaven initiation, one first experiences

the Greater Silence in his skull or cranium directly above Quelle, or the Center Q, residing at the base of the brain. Directly over Quelle, the Father in Heaven makes His Presence known to him who has earned the Shekinah or Melchisedec powers in the Second Heaven.

When one enters the Presence of the Father, he finds himself centered within a deep, amethyst, velvet-like darkness of uttermost quiet. Next, he experiences an engulfment of the Father's Love. And, in sequences of initiations, he is initiated into the Father's Presence, His Light, His Life, His Will.

Until one unites with the Father, he is incapable of knowing the Father as an actual sustaining Presence. All he knows of the Father is through faith in Him, love of Him, and brief or fleeting revelations concerning Him. When one finally absorbs all that the Father may tell him of His Love, all that he may know of Light, all that he may understand of Life—his own will is then ready to be used as an instrument for the Father's Will. In no other way will the Father's Will work through him. Only with the Father's Will may one will a person well.

Jesus is the door to actual experience with the Father. In all initiations the will of man must come under the tutelage of both the Son and the Father.

Awareness of the Father is the last breakthrough the initiate must experience before he may truly serve the Father. The Father doing all things through him, the initiate wills to restore life; he wills to open the Archetypes; he wills to manifest, to de-manifest—all through and by the Father's Will in him.

All initiates are taught the dangers of the fiery trials of the will. The mental will and emotional will are fire; overextension of the self-will harms, burns, destroys. The Father's Will—centered in the will of an adept, initiate or apostle—soothes, heals, resurrects, re-creates.

Should one seek to will or burn aside the veil between him and the Holy of Holies of the Third Heaven, the result is destruction and death to the spiritual life; the raw fire of force minus spirit is a devourer of mind and body.

The first manifestation of the Father's Presence is a peace beyond understanding. The second manifestation is love, a reassuring joy, a praise happening and quickening. The third manifestation is illumination, vision, revelation. The fourth manifestation is a vigor in the body, heart, mind; a vitality and lightness of the body. The fifth manifestation of the Father is spiritual power. The sixth is a healing and restoring. The seventh manifestation of the Father is the power of Omnipresence or the spiritual gift of etheric visitation.

> *No door is closed to the seeker, if he intensely desireth and so loveth that he will continue in his search. The path will open to his feet, and he will find entry into that which he seeketh. He hath been promised the Kingdom, but it is through effort that he findeth it and liveth it. It is God's Will that he seek it. It is through God's Love that he findeth the Way and liveth it.*

THE TELEPATHIC DISCIPLE

> *Cosmos Disciples are the most highly-evolved persons on earth, and at all times are in direct alignment with the Archetypal World or Kingdom of God. Upon the shoulders of these 33 sacred men rests the responsibility of telepathic co-ordination and mediation among the 13 Telepathic Disciples, the 3,000 advanced disciples, and the 300 Good Brothers living in the physical world.*

The great or master images are sealed within the Greater Archetypes. Only a Telepathic Disciple, a Cosmos Disciple, or the Elect may receive directly from the master images within the Greater Archetypes. The task of spiritual revealing is hard for the Telepathic Disicple and Cosmos Disciple, for so many in the earth being more at home within partial truths, fail to recognize new revelation.

A Telepathic Disciple of the mind is one thing and a Telepathic Disciple in Light is another. The Telepathic Disciple of the mind works with men and the minds of men, while the Telepathic Disciple in Light works with the Christ-Mind Archetypes and the World of God.

When the disciple becomes a Telepathic Disciple in Light, he has mastered the forty-nine values of the seven Light-Streams of the Masters. He has a logos of seven spiritual languages, and he is linguistic with forty-nine nuances. Thus, the Telepathic Disciple in Light touches all planes, all spheres, all Realms and all Kingdoms, and interprets them to the mind of him who will hear, see and know.

The Telepathic Disciple of the mind touches the level of the Saints in the First Heaven. He works to overcome the illusion and the fantasy in the thoughts of men. When he has overcome the mist of mysticism, he becomes a Telepathic Disciple in Light.

The Telepathic Disciple in Light learns the difference between choice and decision, intuition and guidance. In choice, one becomes the victim of the perpetual alternates; in decision, one works with the spiritual will.

The disciple comes to know the spiritual will through complete surrender. The will is then given back to him as strength, as courage. From the spiritual will he is given a knowing above that which man knows and the courage to stand and fulfill.

The love which moves through the disciple's emotional

body within the Master's Medallion brings to him the peace of heart and brings to him his Beloved, or Higher Self; but that of love first affects the disciple through men's love-thoughts colored by their passional magnetisms and their negations in love—for the disciple must learn the difference between the passional magnetisms and the pure spiritual love. When the disciple loves purely, he is not repulsed by the moving magnetism of men's love, but he knows the Real in love, and it comes to him as a bird coming home to its nest. In his recallings from other days, when he reads the records of his soul, the first memory he recalls is the memory of love. His desire for love and his hunger for love bring to him the first revealing of his soul's records.

The Telepathic Disciple in Light goes in and out of the Spiritual Worlds. He gathers his recordings, his records. He builds upon the Scrolls in Light. He balances the realities of the physical world with the Real of the Spiritual Worlds. He attracts to himself many who have partial wisdoms and partial knowings. He awaits his time, and says that he knows not all—but he knows that all do their part.

He recognizes the symbol in the forehead of him who also walks in the swift pacing of the Master's Medallion, for such an one also says little and does much in the Master's Light. He says little, but he sings a song—and that song is in the medallion of that disciple seen by the Telepathic Disciple.

God as Spirit projects His Spirit as Omnipresence. Omnipresence is manifested through God-willed minds. The true telepathic filigree supporting the Great Intelligible Unconscious, or Purusha, is inhabited by the countless mind or mental hierarchy-nature vitalities of Great Souls who have lived in the Earth as *Knowers*.

The Greater Telepathies are open to the Telepathic Disciple especially when he is close to becoming a Cosmos Disciple and therefore a direct revealer of Archetypal Light. The Greater Telepathies are those that keep alight the

Light-chain of ideas, passing them onto enlightened souls with hierarchy-nature minds. A Telepathic Disciple has mastered the sensitivities of the senses on the gravity level and is ready to be a vehicle for the Cosmos-Universal Power of Passing.

> *When man hath acquired the means by which he maketh the passage, he will make the journey into the Spheres of Light and recover the tones ringing true to his soul-tone. The disciple in light maketh his passage into the Realm of Light when he hath opened his eternal and spiritual atoms.*

The light in the spiritual atoms of the thirteen Telepathic Disciples and things seen within the great Archetypal Worlds are recorded within a ventricle of the brain heretofore not used by man. This ventricle for the receiving of telepathy will become more sensitive in the scientific age. The Telepathic Disciple has opened a lighted passage in the skull which enables him to record and to remember the instruction he receives from the Archetypal Worlds.

He who hath Logos will speak of the Archetypal Ideas and write of them. He who is scientific will bring them as mercy-inventions to the world. He who hath art will bring them as creation. He who hath music will bring them as masterful composition.

It is the Telepathic Disciple's work to build the body of luminosity and to recall the Archetypes in their order and to transpose them to men as ideas, that men may begin a new form of creation.

The Telepathic Disciple lives always in the knowledge of other disciples and their work in earth. He lives in the knowledge that he is accompanied by other disciples throughout the earth. And he lives in the knowledge that separateness between disciples is one of the greatest sins, and that he who separates himself breaks the link between disciples. He knows that the blending in discipleship is as necessary as air to breathe. He who has the power of discipleship and cuts the cable between himself and another disciple has affected all disciples. The disciple lives and breathes and moves in light, that he may keep forever the tone of blending with all disciples.

The disciple increases his stature from life to life. Through creative embodiments, he increases the light within his body of light.

The Telepathic Disciple looks back and sees that he has been the lesser disciple, the lay disciple, the evangelistic disciple, the patriarch disciple, the prophetic disciple, the disciple of light. And he sees that in time he will become the Cosmos Disciple.

The Telepathic Disciple stands in simplicities, but also in grandeur. The Earth becomes a small globe in the palm of his hand; he discerns the Greater Archetypes. He relates himself to the heart with pity and uttermost tenderness. And he sees all of its convulsions and its evolvings, and all of the great spiritual participants working with the earth. He sees mighty and wondrous things unfolded and revealed through the light of telepathy. And he knows that the minds of men swim as in a Saturn sleep, covered by a sulphurous, foglike cloud over the mental body. The disciple sees that the cloud now begins its dispersing and that light comes swiftly onto man. He who lives within the mental cloud will receive the light as a lightning blast—and he will begin to see other degrees of light rather than his own egotistical light.

It is through the angelic stream that the luminosity body of the Telepathic Disciple is enabled to rise. Through the angelic accompanying, he is enabled to look on the Archangels' work with the planets and with the fiery core of the Earth. He beholds the working of the Race Angels. He sees how nations and races evolve through karmic timings. These things are shown to the disciple through the Angels and the Archangels.

Night-flight for the Telepathic Disciple becomes a matter of organized ether and organized light. The design which he carries out in the day comes from the plans made with his Higher Self in the night. Through grace he has earned the right to rise into the higher Spheres and Realms of Light.

Karma and entanglement remain not long in the life of the Telepathic Disciple. The Telepathic Disciple rises above the entanglements and disengages himself, that he may work in the right harmony and the right environment.

There are no exceptions in karma for the Telepathic Disciple. He pays his karmic debts with the unique dividends of grace; and he finds strength to disengage himself from that which restrains him.

The teachings and patterns of Heaven blend in man's outer consciousness, for each thing is interpreted after his own expression. The Archetypes of Heaven, on reaching the brain of man, are sifted through many levels of consciousness, and appear to him after his own interpretation and manifestation. Slowly, men are acquiring and adding to their knowledge the experience in the Spiritual Realms and the memory of Beauty, Love and Light.

SYMBOLS

All sentient symbols stem from the lower astral world, the higher astral world and the reflected archetypes, and are contacted through an etheric braille. All initiatory symbols stem from the Spheres of Light. All superconsciousness symbols stem from the Greater Archetypes.

In the first four Great Intervals of this eternity system, our Father—working with the Elohim-Hierarchy and the Archangelic Presences—imaged, formed and shaped all things. The Father established the original Blueprints or Archetypes for each thing imaged. Within each Archetype He implanted a *master image* containing multitudinous symbols.

After the four Great Intervals of the *forming* period for this eternity or earth system had concluded, the Father gave the earth and its contents over to the creative faculties of man. Thus began the *informing* stage, which will last for the three remaining Great Intervals of the world. In the informing stage, man is informed or instructed.

Though one may be continually the recipient of creative Archetypal formats and symbols, he must earn the right, through certain initiatory trials, to interpret and to ethically use them.

All Telepathic Initiates are in alignment with one or more master images. These master images are sustained within the Hum of a Cosmos Archetype. In this eternity system there are thirteen basic Cosmos Archetypes. Within the thirteen basic Archetypes, a master image remains for ages or even for aeons. The master image sends forth billions of symbolic codes or etheric forms. These eventually fall into the receptive minds of those who creatively interpret and activate them.

All the words written in the earth of Sacred Scripture have been built first out of seven symbols. These seven symbols go back into the Original Archetypes and are reassembled onto the things of Light and the things of Good.

Symbols are living Archetypal particles of light that never change in the inner consciousness. The higher unconscious is familiar with symbols. One is constantly absorbing symbolic nuances into his outer consciousness.

All persons constantly use symbols in their expression. All

thinking and feeling are dependent upon symbolic life. The emotions are especially dependent upon the energy symbols emanate. The mind depends upon the meanings within symbols. Symbols are stabilizing reassurances from birth to life, from life to death, and from death to birth.

Symbols are the vocabulary in dream life. In the life between death and rebirth, the language used is the language of symbols. God has given man symbols, that he might extend his powers of concentration and use the energies of his mind to create, and that he might move over the limitation of language, race and place.

There are seven great master symbols centered within seven great Master Archetypes. When one penetrates the master symbol in a Master Archetype, he opens 144,000 symbols. These are released into his consciousness. From this he gathers an amazing inner vocabulary of the mind.

When the symbology of a tree appears in a dream, the first symbolic level of the tree is Archetypal; such a dream means that one either has been given instruction about the Greater Archetypes or he has observed the working of the Archetypes. When one sees the top of the tree, he is observing man's mentality. When he sees the limbs of the tree and their branches, he is relating himself to a Family-Atom. When he sees the root of the tree, he is researching the memory of the ages.

In the higher astral planes, trees may be seen sometimes in an upside-down chimera; that is, the roots appearing at the top and the foliage at the bottom. This indicates that one is viewing an earth archetypal-tree relationship, that is, the earth is receiving power from an Archetype.

> *The Lord said: They speak of an imperishable Asvattha Tree with its root above and branches below. Its leaves are the Vedas, and he who knows it knows the Vedas. Above and below spread its*

> *branches, nourished by the gunas. Sense-objects are its buds; and its clustering roots spread downward in the world of men, giving rise to action.*
> —*Bhagavad Gita*

All symbols correlate to the Greater Archetypes and are of etheric origin. All symbols are energized vortices containing a life ether and a light ether. The life energy and ether mold and shape the form of the symbol, and the light ether is responsible for sustaining the intelligence speaking within the symbol.

As long as man has a possessive desire of any kind, his symbols will be colored with something of the sentient. When he is wholly selfless, he will receive the true superconsciousness symbols, and thus become at one with the Christ Mind. The true and pure consciousness symbols may be received only when one is in direct alignment with the master symbols of the Greater Archetypes. *A perfect consciousness symbol is an Archetypal idea which overcomes the barriers of Time and Space; Heaven and earth are one in a perfect consciousness symbol.*

Few persons exist in the world who can open the Archetypal symbols and free men into the full-energy mental and emotional capacities of their souls.

To be a real and true Teacher, one must have access to the constant wisdom-consciousness stream of parabolic allegories and the Archetypal symbols. Otherwise one is a parrot or imitator and is devoid of the Living Word within his speakings or sayings.

All symbols are, in reality, deathless, immortal, living nuclei. Centered within each symbol is an etheric, intelligible code. In each eternity, symbols play their part to make known to all conscious beings the Archetypal Blueprints and intent of God.

Regardless of what language is spoken, the cosmic symbols in this eternity enable men to retain something of an inner, intelligible communication. The hidden meaning within symbols holds the key to the inner understanding between men.

> *The ascended arc is the rainbow symbol of mediative harmony. May I blend with each perfect prism, and become at one with the many facets of the Perfect Self.*

16

THE GREATER ARCHETYPES

Everything in this Earth comes as an idea which has its origin within the Greater Archetypal Energies within the Mind of God.

Archetypes and Archetones

In the beginning was the Word, and the Word was with God, and the Word was God.

—St. John 1:1

The Archetypal Atma or the *Word* is the Central and First Atom holding together all Galaxies, all Suns, all Earth-systems, all Creations. The Archetypal Atma is the Godhead from which all Energy and Vibration stem as Intelligible Consciousness.

The Archetypal Intelligible is *God.* God as Soul and Spirit is Godhead in man. By rates and measures, men record the rotation and movement of worlds within worlds, whereby the Intelligible makes known His Manifestation as Consciousness in Man.

When this Earth and its accompanying planets were established, a Greater Blueprint or an Archetypal Envelope was placed around the eternity system.

One's own Archetypal Image is the Image of God within himself. Each life is given to one that he may reclaim the virtues gained in former lives and use them to bring forth latent virtues which are potential within the Archetype or Image of God within him.

An *Archetype* is an Original or Divine Blueprint for the mind and life processes in the Earth.

The Virginal Archetypes are the Unmanifested Blueprints or Ideas of God awaiting the right timing to come forth into the consciousness of man.

The Greater Archetypes work first as the Unmanifest or as the Word yet to be made flesh. These spiritual, creative, divine Archetypal compulsions move in tides into the world through the minds of men.

The Greater Archetypes concern the greater life-waves directing and changing reincarnation tidal flow. They also determine the timing of the appearance of new ideas in the receptive minds of the Elect or Adepts in preparation for new eras in time and evolvement.

All true ideas have their source or origin in the Greater Blueprints or Archetypes of God. God's Equation determines when men are ready to receive the greater truths. Man's degree of evolvement determines how he will interpret them and use them.

The Archetypes are quickened by the Father, the Hierarchy and the Christ. From time to time, men receive these great imprints upon their thoughts. Where men are more materialistic in the earth, they are unable to receive the reflections of the Archetypes.

Those who live close to the Greater Archetypes of manifestation-ideas are our teachers, our peers. The initiate prays for spiritual worth rather than for physical wealth, as deep in his heart he knows with a sacred and divine intuition that the lasting and true substance begins with the recognition of God as idea, God as Creator within all ideas.

In each Great Archetype is an *Archetone*. The Holy Spirit sounds within the Archetone when the Divine Word is ready to enter into the mind and life of man. When the Archetone sounds, the Archetype comes alive with whatever compulsion is ripe or ready to be manifested. Thus, an Archetype remains dormant until the Archetone sounds.

The most deep desire of all initiates is to contact the Audible Life-Stream, or the Archetones upholding all life. The Audible Life-Stream is the Life Current, the Sound Current, the Spiritual Current of Logos flowing forth from the Greater Archetypes. Christ has command over the Archetypes supporting the Mind of God.

The *Word* spoken of in the Bible is the Vibrationless Vibration or the Soundless Sound. To unite with the Soundless Sound or the Sound Current in its vibrationless state is to unite with the unconditioned Supreme One, or God. To *hear* the Audible Life-Stream is to unite with the conditioned God, or our Father which art in Heaven. In both waking and in dreams, one unites to some degree with the Sound Current of the Celestial.

The Being called the Holy Ghost frees the Sound Current of the Great Archetones. The Holy Ghost working with the Will of God sounds the Archetone that the great ideas or new impulse for man may enter the world or fall into the mind of man in timing to the Will of God.

Christ is the Diamond Center or Central Point in the Archetypal Realm. Jesus' dedication as the Son of Man in the world enabled the Christ to polarize the souls of men in earth.

With the coming of the New Day or the Archetypal Power being given in this present time, those who have carried throughout the ages the Gifts and the Powers of the Dharma are provided with the Ethic in the use of Enlightenment-Teachings.

Jesus came to plant the Archetypal Seed for the use and revelation of the mysteries or the meaning of man's life in earth and what is required of him who contains the Holy Truth.

> *During the birth of the world, God—the Eternal Spirit—did plant the greater Seeds or Archetypes. These mighty Atom-Seeds or Archetypes now shower upon men. All who have hearts of purity, all who have given of love, shall be taken into the net of the good Fisherman, the Lord Jesus—and the human spirit shall come forth. The human spirit shall work through those who serve in naive good. These now enter into an aroma of preciousness, expressing the true human spirit.*

When Saint Paul heard "unspeakable words" in his initiation in the Third Heaven, he heard the Archetones in the Greater Archetypes and knew no words in human language by which he could interpret to the unknowing mind that which he heard. He had touched the Greater Unmanifested which man was not yet prepared to receive. This is so in the life of all persons uniting with the Audible Sound or the Holy-Ghost powers. Unless one has earned the three powers of the adept, he cannot always articulate that which he hears. Even so, he cannot always give forth what he hears to unripe or unready ears. The three powers are true seeing, true hearing and true speaking. To achieve this, one must have reached Timelessness in his evolvement.

Through union with the Greater Archetones, the seer of hearing reaches the very core of union with God. He unites with the Hum or the great Alpha and Omega Tones working with Universal Cause. He moves out of the causal time limitations of the physical world, and thus experiences Universal Cause.

All seer-initiates work to remain in union with and make clear the Audible Sound. The Audible Sound Current can be felt as an outgoing, expanding, pulsing, ecstatic joy emanating from the eyebrow center. When the triune powers of seeing, hearing and speaking are activated, the physical senses are muted and made quiescent. The lower aspects of the senses become as lowered flames in a lamp, while the innermost light of the mind sees, hears, and speaks of the Real.

True inner hearing has not been perfected until one can hear the sound in a seed, in a leaf, in a tree, in a grain of sand, in a flower, in its fruit, in a mountain, in a rainbow, in a drop of rain. When one unites with the Archetone giving life to all forms of life, he unites with the Cosmic Music of the Universe. Finally, one unites with the tone of his own direct star from whence he came.

Music begins as tone in the great Hum or in the WORD sounding through the Greater Archetypes, from whence it is sent down due to the gravity pulls of the physical earth where it falls upon the astral nadis centered in the astral core around the spinal canal. Here, as vibration, it flows into the aura and then into the oral etheric vortices surrounding the ears. Vibrational flow transmitting music as sound, harmony and melody is heard by the hearer as to what *he is* in his evolvement. No two persons hear in the same manner. The nuances of sound cannot become more than *one is*.

In the coming ages, language as it is known today will become obsolete. There will be a new tongue or a new vowel uniting all languages. In the present time, music is the common language for man.

Holy Divination is in everything when the Eye is open. Holy Divination has a language; it is in the Holy Writ of the Archetype and in the Holy Witness of the Archetone. Holy Divination as speech is the Hearing of the Word and the Healer Vehicle for the Soul.

The Living Teachers are living books having access to the

Word through the Greater Archetypes. Besides these, there are lesser teachers not having access to the Greater Archetypes; they are partial-truth purveyors of the Word. To fall into the hands of a biased teacher is to feed upon the husks of unreality.

The greatest thing a Teacher may pass to his students is the *Esse. The Esse is the Essence of God overcoming absence from God.* The Teacher who passes the Esse and the Ethic to the student is the Teacher having the grand elegance or master elegance of the Archetypal Logos.

> *The source of the Parabolic Chain exists in the Archetypal Realms.*

The *Sacred Name* is the primordial identity name. This name dwells in the Great Unconscious of the Archetypal and Archetonal vibrations encased in the Constant.

The Great Archetypes and Archetones in the upper reaches of the Third Heaven have a level of sound which contains the Name-of-God Mantra. The Name-of-God Mantra as the Constant over all life is an Archetypal vibrational tone sounding into the Constant of all consciousness beings in the earth. To receive a *Grace Name* is to connect oneself with his Constant and his own Sacred-Name sound in the Great Unconscious of the Third Heaven.

When one is free to see with holy vision into greater dimensions of light, he sees beyond the personal or partial, and he will eventually receive the power to look upon the Greater Archetypes or Blueprints of God. He beholds the Plan of God and His finished work.

> *It is through the downpouring of the Greater Archetypes that he who has mislaid or slept to the Commandments must retrace his steps and call them forth from his sleeping state of conscience.*

ARCHETYPAL CYCLES

God builds everything in order through Cycles and Rhythms.

In Omnipresence cyclic Law, God as Law reproduces Himself in the human realm by and through cyclic repetition. Perfection abides within the Law of God's Increase and expresses itself as perfected consciousness.

When a shift in Galaxy occurs, the Great Unconscious simultaneously produces an Archetypal ripeness. This Archetypal ripeness changes the gravity density obstructing soul and outer consciousness.

A shift in Galaxy produces an *Archetypal Cycle*. During this time, there is a greater expansion in spiritual enlightenment and revelation. Such periods wipe out ignorance, and may be compared to the rising of the Universal Sun, whereby the Universe and man have entered another Day, a Day of expanded consciousness. In these times, the Archetypes distribute the Great Light falling into the minds and lives of those ready for enlightenment.

Accompanying each Archetypal Cycle and Galaxy shift is the changing of the rotational theme in planetary bodies. The obscuring primal-veils veiling away planetary knowledge are lifted; man opens interterrestrial knowledge regarding the Solar System he inhabits.

There are miniature cycles, major cycles and mighty cycles. The Archetypal cyclic flow works with all of these. New and virginal Archetypal beginnings are the result of miniature, major and mighty cycles coinciding or working simultaneously to bring birth to more in consciousness. In no other way may man leap over the scales of the units of measure.

Man in Maya is organized to live an enclosure existence. Freedom from these enclosures, which confine him to the belief in limitation, must come upon him involuntarily as being outside of himself or his own control.

If one lives consciously within the Archetypes, he is in the Great Rhythm; his cycles move faster, and any alien body is discarded and removed. In the Great Rhythm, there are Cycles of which men are unaware. In their tamasic and rajasic natures, they cannot register the velocity of these Cycles.

During Universal initiatory cycles for the masses of peoples, Omnipresence, through the Divine-Companion Presences, moves the seer or prophet beyond the gross cycles. The seer is attuned to the Omni-Dimensional Cycles which are the refined Intelligible, Universal Creations. This is what happens when one becomes a prophet for the Archangels.

One must come into rhythm with the Universal to master the Cycles. Through devotional and spiritual practices, one takes hold of the rhythm controlling the greater Cycles. By this he frees his Cycles and lives in his true units of measure.

> *Everything one does activates a Cycle or a correspondence to Cyclic Law. Every Cycle is as a seed carrying its own Archetypal compulsion to expand.*

THE THREE ARCHETYPAL ACTIONS

> *There are three Archetypal actions affecting the earth: (1) the Greater Archetypes, (2) the Moving Archetypes, and (3) the Destroying or Sealed-in Archetypes. When the three Archetypal systems coincide in their actions, men in the world experience emotional, mental and religious chaos. The outer friction caused by this combined spiritual fire purifies the masses, and prepares for a new era in which men will aspire once more to come nigh to their souls.*

The Greater Archetypes in the Third Heaven are throbbing, humming vortices of intelligence. These Archetypes are the connecting link between the outer universe and the universe of man.

The Greater Archetypes are opened to holy seers only when men on earth are prepared to receive eternal and spiritual ideas.

The Archangels working with the Christ preside over the Greater Archetypes. The Christ Spirit is the Archetypal Presence permeating the Greater Archetypes. When the Telepathic Disciple unites himself with a composite telepathy in the Third Heaven, he is taught directly by what he sees and hears in the Greater Archetypes.

He who rises to the world of the Greater Archetypes communes with the Archangels overdwelling and protecting the Archetypes. Through the help of the Archangels, he sees the Eternal and Eternity Blueprints, and he records them upon his mind.

The Greater Archetypes work with the *Will of God*. The Archangels sustain and maintain the rhythms of the Archetypes. When men are ready to receive the greater ideas of Light, the Holy Ghost releases the Tones or Archetones within the Archetypes—and saviours, holy seers and prophets transpose and reveal the ideas of Light to men.

The Greater Archetypes form a cohesive body, enabling the Christ to center His Light in this earth system. The world is now undergoing a cleansing so that men may prepare to receive the Archetypal Light of the Christ. Institutions, barriers and broken idols will be replaced by the true fire of the Living God through which the Christ, the Regent of the earth, makes His Presence known to men.

When men are reluctant to believe in God, the world falls into a tumultuous chaos. During a chaos period, a nucleus of advanced disciples is born to the earth. These disciples know one another in the Spiritual Worlds. They have made a covenant and a vow, and they are protected from the forces of the dark. Their work is to bring the Word of the Kingdom of God; and from their teachings the hearts of men are made more humble.

The *Moving Archetypes* dwelling within the Second Heaven are three-dimensional reflections of the Greater Archetypes. These Archetypes work with the Laws of God. The Moving Archetypes move upon men every ten thousand years, and stir them to form new religions; to transform racial impulses; to build societies; to support the moral ethic protecting families, communities and religions; and to produce ingenious inventions suitable to the era or the age.

The *Destroying* or *Sealed-in Archetypes* reside in the core of the earth. These Archetypes are absorbed by men into their emotions when men become decadent, agnostic, corrupt. The Destroying Archetypes tear down old systems and formulas so that the Greater Archetypes may touch the minds and souls of men in the world.

The Destroying or Sealed-in Archetypes are reflected in the lesser regions of the astral world. When men see these Archetypes, they "know in part"; they come under the fear of the wrath of God; their prophecies and their dreams are colored by subtle fears.

When men come under curses, they are under the influence of the Destroying or Sealed-in Archetypes. Only their Guardian Angels may free them from the devitalizing conditions invoked through harmful or cursing thoughts.

Before mankind as a whole can receive the spiritual impulses sent forth from the Greater Archetypes, certain egos are born to the world—egos whose purpose is to work as ambassadors for the Destroying or Sealed-in Archetypes. Having lived many negative lives, they become karmic pawns in society. Their negative, destroying tendencies make them ideal tyrants for destruction. When old systems are to be erased, rulers with agnostic hearts feel compelled to overthrow governments, societies and religions.

When glacial periods and earth cataclysms—such as earthquakes, floods, volcanic eruptions and tidal waves—occur, the Sealed-in or Destroying Archetypes are at work. During these times men die, that they may be reborn in periods more suitable for their evolvement. Those who survive are shepherded or herded into continents waiting to know the hand and skills of man.

The Greater Archetypes occur in 250,000-year intervals. In this age, the Greater Archetypes, the Moving Archetypes and the Destroying Archetypes sound their Tones simultaneously, and men of the earth are being created in the vision of the Christ—that is, given new birth-impulses within their souls.

The Archetypal Kingdom is the Great Unconscious of the Unmanifest, or things to come. On the level of man, the Archetypal Kingdom functions in periods and cycles of 12,000 years. In 10,000 years of these, man is initiated through whatever Saviour provision is given to discipline him and guide him. This 10,000-year period is called a *Moving Archetype*.

The 2000 ending years of the 12,000-year period are interim years through which mankind makes union with God without resistance. In these interim 2000 years, whatever degree of faith one has is made perfect for him; he has a soul-respite from the Maya rajasic pressures and tamasic inertias. During this time, that which determines what he will do through reincarnation will imprint itself upon him as Destiny for the next reincarnation cycles in a virginal Moving-Archetypal timing or time. This is what the Scripture means by "the sheep and the goats." Man's own virtue or non-virtue places him into reincarnation polarities and functions where he fits or belongs. Always in man, however, is the promise of the Risen Christ and his hierarchy nature coming to birth.

Every 10,000 years, man enters an Archetypal action.

This 10,000-year cycle is divided into four periods of 2,500 years each. In the earlier phase of a 10,000 year Archetypal cycle, men are established as to moral laws, as to race, society and nation. In the last phase of a 10,000-year Archetypal Cycle, men enter a spiritual destiny. When this has been concluded, men begin again upon another 10,000 year Archetypal Cycle, retaining the essence of good which has been established within the former period of time.

The Archetypal Cycle in which men in the world are now involved began with Abraham, and shall be concluded 5,000 years after the birth of the Lord Jesus. There are certain periods within all Archetypal Cycles in which men lose the thread of their oneness with God and fall into materiality. While in these periods, they produce vast civilizations; then follows a period of corruption. And, after this period of corruption, the helps of Mediation from the Higher Worlds enable men once more to rise.

The end of a 10,000-year Archetypal Cycle is called a Latter Day.

At the conclusion of each Moving Archetypal Cycle, certain persons who refuse to evolve and to move with the tide of their times as to its spiritual portent, and those who offend the greater laws, become stragglers, and are withdrawn into a twilight state. This occurs at the end of each 10,000 years. This twilight state is in reality the 8th sphere, and is spoken of by Jesus as "outer darkness." It is known by all initiates that in certain levels of the 8th sphere there are some who believe they still live within the earth.

All spiritual evolvement is overdirected by the downpouring of the Greater Archetypes, the Moving Archetypes and the Destroying Archetypes—all synchronized together to thrust men forward into a great New Age where men will know God and receive directly as persons the first and pure Archetypal commands.

He who receives the Greater Archetypes is aware of the

Christ Spirit as being the only begotten Son of God. He recognizes Jesus as the Son of Man who as Christ Jesus is the Sovereign Channel for the Greater Archetypes.

In a synchronized period of the Archetypes such as Enlightened Souls are experiencing, Jesus is the center of Mediative Light. He is the Path. His Ethic, as given in Saint Matthew 5–8, is the Way.

Through ethical procedures, one opens himself to the holy audibles or to the resounding of Jesus' Words, to the reproduction of His Miracles. In one's own behavior and conduct, he lives the spiritual life, and advances in the expanding consciousness state through initiatory trials and processes which duplicate to some degree that which was laid down by Jesus as He walked the earth and by His twelve disciples who inherited the Word of God through Jesus.

When men come to the time in which they will relate themselves to the Greater Archetypes, the Moving Archetypes and the Sealed-in Archetypes, they will have perfected the greatest system in mathematics; their understanding will be through the four-fold consciousness; they will have penetrated Space; and they will have mastered Relativity and Time. They then will have come to the point of the spiritual equation working within the Christ, the Father and God.

THE MOVING ARCHETYPES

> *There is no master key made which preventeth the spiritual side of man from receiving his answer and opening the door. There is no design created which doth not achieve its purpose, if it stems from the Archetypal Design. There can be no failure if that which is conceived in the beginning is out of the True and the Whole.*

The Mathematics of the Cycles enable one to understand the lesser and the greater Archetypal Days. There are periodic impulses which determine the Genesis states of man in his evolutionary processes. The Mathematics of Cycles lead man directly to Universal Mathematics whereby he unites with rhythmic compulsions directly sent forth from God as Spirit.

The rhythmic impulses relate to the Greater-Archetypal Days; the cyclic impulses, to the Moving-Archetypal Days. The Greater-Archetypal Days occur every 250,000 years. The Moving-Archetypal Cyclic days occur every 12,000 years.

Four Genesis-Cycles in a Moving Archetype repeat themselves in each Moving-Archetypal Cycle, always seeking to become more refined and sensitized. Each Moving Archetype has four *Yugas* or Ages.

At the end of a 10,000-year period in a Moving Archetype, there is a 2,000-year level of God-knowing. God-knowing, regardless of the level of knowing in a person, relates to three things: (1) knowing of existence, (2) knowing of knowledge, and (3) knowing of God. In a 2,000-year Archetypal-Interim period, those who are less evolved will know God only as existence, as life. Those who are Niscient or *Knowing* will know God directly through Soul-experience. Those who know God as Bliss will achieve Samadhi, and will be of those 144,000 who will enter the Saviour Path or Bodhisattva Path to assist all others in the world to achieve God-knowing.

In every 12,000-year Moving Archetype, during the 2,000-year Interim time those who are ready to enter into the true Bodhisattva enlightenment become the spiritual avatars for the next Moving Archetype.

There are periods in which the Greater Archetypes are more active than in others and more effulgent with the

Word. When the world is ripe or ready to receive the greater ideas from the Third Heaven, the Holy Ghost speaks through the exalted Tones or higher Archetones within the Greater Archetypes. When the Greater Archetypes are dormant, men must depend upon the Moving Archetypes for their religious and moral impulses.

Jesus came in the middle of a Moving Archetype, preparing men for spiritual birth. At the ending of each Moving Archetype, men are graded as to souls and degrees of consciousness.

In each Moving Archetype, men are on trial that they may develop certain stages of Genesis-evolvement. Reincarnation-evolvement determines what Genesis one expresses in a Moving Archetype. In each Moving-Archetypal cyclic portion of Time and evolvement, men repeat Tribal-Genesis, Family-Genesis, Lower and Higher Self-Genesis, and Cosmos-Genesis.

When 10,000 years have concluded in a Moving Archetype, men experience an opening to God. When one is in Tribal-Genesis, he experiences God through Nature. If one is in Family-Genesis, he experiences God through a Family-Atom, through blood relationships and through marriage. If one is in Lower Self-Genesis, he magnifies his ego. If he is in Higher Self-Genesis, he spiritualizes his intellect, seeing God as the One in All. If he has reached Cosmos-Genesis, he is a Cosmos-Disciple, and he has established a likeness of John the Beloved, the closest disciple to Jesus.

At the end of every Moving-Archetypal Interval, men return again to the shaping and forming process of a new Moving Archetype. In the present time, every man in the earth has something of Tribal-Genesis, Family Genesis and Self-Genesis in his nature. This always occurs in magnified form at the ending of a Moving-Archetypal period.

At the end of each Moving-Archetypal period, the four Genesis-attributes in mankind experience an alchemization. When a Greater Archetype is ready for manifestation, it sounds its tones upon mankind; there is mighty chaos in the world regarding clans, tribes, clusters of men, families —and the individualistic nature of the self is searched as to ego-integrity, ego-value.

God repeats Himself through rhythms. God activates Himself through cycles. The Greater Archetypes work in rhythms; the Moving Archetypes, through cycles.

In the present time, the great rhythmic flow of the Archetypal Light is conjoined with the Moving-Archetypal energy and with the Destroying Archetypes which work to eliminate, to assess and to judge where one is, what he is, and how he can be used by the Power of God.

The Archetypal nature in man seeks to perfect the *hierarchy nature* in man, that he may move into his heritage as a willing and mind hierarchy nature.

God is not far off. He is *Omnipresence* in His rhythms and His cycles. The Archetypal rhythms work without ceasing within the Law of God. In the Moving Archetypal cyclic states, man knows this Law and experiences it as the Law of Karma.

In the present epoch, or ending of a Moving Archetype, men are experiencing an overlap between their unknowing darkened past and their knowing or Niscience vision for the future. Men are now in stress, and for the next 3000 years will be in a state of tension caused by the equipoled coordinating processes of the three in the conjoined Archetypal Flow.

Over and over, men are reincarnated that they may realize their hierarchy natures to the fullest, and thus become creators in mastery of Earth and planetary forces.

All Scriptures of the past relate to the Science of Practice. All Scriptures of the past point the way or the method to reach perfection as laid down by procedures stemming from Wisdom-related souls, hearts and minds. *Nothing lives Scripturally unless it contains lasting truth based upon the Spirit of Truth.* The Scriptures are therefore preserved from era to era or from one Moving Archetype to another.

As men move toward the ending of the present Moving Archetype, called by certain schools in India the Dwapara Age, men will become more and more aware of the need to spiritualize their intellects.

During this era, which men intuit inwardly, the Sun of this eternity system and its planets will enter into another Archetonal beat of energy processes. Nature, as men know it, and physical nature which man possesses, will move into other gradations of refinement and development. These occurrences will begin at the ending of this present Moving Archetype, which will be 5,000 years after the birth of Jesus.

In the world today there are but 3,000 advanced disciples. They now prepare to spread their Light.

The Lords of Karma work with the greater Recording Angels. At the end of each 10,000-year Moving Archetype, they weigh and judge the intent of men.

The Lords of Venus work with the Cherubim Angels to enable men to unite their genius with mightier creative skills.

The Lords of the Flame are the Seraphim or Lords of Mind who work with the Christ Mind and the Greater Archetypes, that men may become mental creators or manifestors for God.

The Destroying Archetypes

Those who are sattvic will rise to the Increase of God through Christ; as peacemakers, reconcilers, they will be tried first in the school of Intentional Suffering through Initiation. They will, while being initiated, bring forth through their divine natures the Archetypal Truths.

In any eternity system, when the Cycles and the Rhythms

of the Cosmos come to equal timing, the Archetypal explosion is inevitable. Mutation on all levels of the earth and in the earth occurs; the Destroying Archetype, working with the Archetone of the Archetypes, sounds the tone for the withdrawal of many forms of life and of life-expression. So it is happening in this period, this time.

Because men cannot see beyond their enclosures, they will suffer more; they suffer in bewilderment. This suffering, which is the cleansing flame produced by the Great Archetones within the Destroying Archetypes, sweeps before man, seemingly taking from his personal volition and initiative. False governments, ineffectual religions, ignorance of ethic—all occur. He who is the Christ Spirit, the only begotten Son of God, has set alive soul and mind tremors in the earth through which men are raised up en masse.

The Christ through Jesus is the Uniting Principle. Jesus through man is the Reconciliation Principle. From these come the religiosity soul-expression through faith in God.

The Destroying Archetype works as the arm of judgment within the Law of Karma. Everything that resists the Omnipresence of God is irritated and worked upon when the Rhythm of the Greater Archetypes coincides with the Cycles of the Moving and Destroying Archetypes. That which is inanimate, tamas and inert is subjected to explosion; and the resisting in the rajasic, during the period of the coinciding of the three Archetypes, discloses in their functions all negative frailties and weaknesses. During this interval of time, advanced initiates and also novitiates of the soul discover unusual techniques, utilizing heretofore unused technologies—thus, science on the kinetic level has come to earth.

The present era is the atomic era in which new technologies will challenge the mental and will proclivities of man, and will recover some of the virginal, primal, instinctual powers from the past.

It is always the case in a great conjoining of the rhythmic and cyclic laws that men seem to be unwittingly cast into a cauldron of chaos. God is always the same, yet He never repeats Himself. God increases at all times, in all periods, epochs and aeons. However, He has given to man free choice which will ultimately become free will. It is not that man is God; it is that man is within God. The Eternal Spirit of God in man gives to him through the extension of free choice the power to differentiate himself. All of the various vehicles God has given to man to live in this eternity system develop themselves through differentiating processes.

The cunning and chicanery which support the present age are not of the Father. These come from the Destroying Archetype, which works when races and continents are to be revolutionized.

The hierarchy-nature imaging power is most successful when one is optimistic. When one is depressed and negative, fearful and unbelieving, he is united with the Destroying Archetypes. One who is continually in a state of depression assures himself of a life of limitation and of failure. All healers are enthusiastic and joyful transcribers of their Archetypal Imaging Power.

The hierarchy nature uses the technique of the *Word made flesh* through Christ Jesus to see the ultimate manifestation that all men be lifted up—and that all men become like Jesus or to *see Him as He is.* One side of the hierarchy nature *uses* its destroying Archetypal powers to wipe away the guilt memories in the Undersoul and to clean out the negating fixities that produce the climactic states or cosmic traumas upon men.

The Destroying Archetypes are regulated to destroy through the Time-system in this eternity system. Man's Time-mechanism is presently being overstimulated by the Luciferic or Satanic Archangel, inspiring people to sit at the feet of the media, or TV, and to be held hypnotically into the dramas of violence. These dramas implant samskara-seeds of evil suggestions and motivations into the receptivity of the present or modern aspect of the subconscious of the individual and of the world.

> *"No man can serve two masters: for either he will hate the one, and love the other; or else he will hold to the one, and despise the other. Ye cannot serve God and mammon." (St. Matthew 6:24) Are you working with and for the Destroying Archetypes or the Christ Archetypes? Which are you serving?*

If a person is mesmerized by the tamasic concepts of life, and putting them into practice in his habit and thinking patterns through the use of drugs, through delegating his responsibilities to those more responsible, he is serving the Destroying Archetypes, which are used by Satan to tempt and weigh man.

The Satanic *principalities, powers* and *rulers of darkness* use the Destroying Archetypes to work through tamasic and lower rajasic persons. The degrading aspects of a society are kept alive by the Destroying Archetypes through which Satan works. His dark kingdom is established upon negativity actions and thoughts.

To be habit-ridden as to any vice which is harmful to body and mind, is to be a slave to the dark and for the dark. Tamasic inertia—when looking upon the dark in others, in environment, and in acceptance of darkened influences—makes one a slave rather than a master.

The only hell existing is created by man. The negative currents of the Destroying Archetypes nourish the grey worlds of unretrieved, unrepentant ones in the world who make themselves agents of the dark. Being blind through self-destruction, they know not the Christ.

To move beyond evil, one must take the first step, i.e.,

service for the good of all rather than that which serves the Destroying Archetypes.

When men become self-satisfied, and contented with their crystallized prejudices, the Destroying Principle is activated in the Higher Worlds; a tone is then resounded upon the emotions of men. This stimulates the pioneer spirit in the world, and, thus, through changing circumstance the mirror of life is enlarged, a new vision is seen.

ARCHETYPAL AND PROTOTYPAL KNOWLEDGE

Man is a prototypal, God-imaged vehicle expressing the Archetypal Light or Mind of God in his soul and within his consciousness. A prototype is a man-being projected and imaged by the Will of God. A prototype is a Divine seed-form or zodiacal mold-pattern vehicle expressing an individualized Archetypal Idea or Image reflecting the Cosmos Mind of God.

By Cyclic Law, one moves upward. By Rhythms, one manifests within the greater Archetypal Knowledge, where he is made God-aware in the state of Universal height, length, depth, breadth.

All knowledge is inherent in the spirit of man. Archetypal Knowledge, the greatest of all, produces bliss containment, bliss support, bliss happiness.

Man is not alone. His life is knowledge.

The Great Unconscious is the *Purusha* or the Archetypal Knowledge. The Great Unconscious is opened to certain advanced souls who have the power to reach and to familiarize themselves with the coordinating and collective creating forces of the Universe. Those having this power bring soul-knowledge to men of the earth before it is proven by science. These foreseeing ones prepare men to receive soul-knowledge directly and to put it into action through events which appear seemingly out of nowhere, but which in reality function as a direct process of revelation and creation.

The collective Higher Unconscious is presently pouring down to mankind of the Earth heretofore unmanifested Archetypal Ideas providing the inspiration for the building of the new human race, a race with one foot on Earth and one foot on outer-space terrain.

Archetypal Knowledge is irresistible. It takes hold of ignorance and through transenergization shreds it into nothingness. With response to Archetypal Knowledge, man must meet destiny with a whole mind which exists within his soul self-nature.

The hierarchy-nature healer works for the world-soul as well as the individual soul. His mental and heart atoms operate on the level of the eternal atoms of the higher unconscious, where the Archetypes exist.

The hierarchy-nature healer's thinking and care for the sick see the illimitable grace provided for all souls in prototypal form. His hierarchy nature abides within a fullness of grace in total awareness of God's good pleasure and love for His sons of the earth and of Heaven.

Man is a prototype. An Archetype is a Blueprint of God, or a nucleus of spiritual ideas. Through repeated lives, each person incorporates into his soul and his ego something of the Archetype of ideas of God.

While the soul remains the same, each life portrays a different prototypal action, and thus extends the range of the soul's action. Eventually, when each person of the earth has incorporated all twelve of the prototypes, he will become a perfected prototype; he will become like the Son of Man, Jesus.

Men must take hold of themselves. Through the inward life, they must grow and observe the growing. Through the inward involvement, they must shape and hew out their own Archetypal and Prototypal themes. God has so willed it that man must conjoin the personalized forces of self and consciousness within his own true existing self with the Power and Omnipresence of God, that he may fulfill that which is destiny for him and this earth—this eternity system of solar light, moon light, planetary light and soul light.

The search for truth begins with self. Soul is ever waiting for the awakening of this hunger, this desire.

The soul is ignited with energy-vibrating joy when one surrenders to God. When man surrenders to God, holy music is resounded in Heaven or that world where dwells the Divine. The seed of soul presses into earth existence to break open the soil of materiality and to thrust one up to the Light to become Light-Realized, God-Realized.

The Polaris energy takes possession of the cells in the brain, quickening the etheric brain so that the Archetypal Images of greater dimensions may remain active; superimposing the prototypal energies, thus giving man prototypal guidance within the Christ Mind.

Prototypal Knowledge coming from man's experience as ego, when used over a period of time, loses the vital archetonal sound vibrations. This causes dogmatic crystallization.

Archetypal Knowledge and Prototypal Knowledge for man are controlled by cycles and rhythmic tides—for the ego, 12,000 years; for the soul and higher mind, 250,000 years.

Prototypal-Archetypal Knowledge shapes the life of the ego. The developing physical and emotional mental energies experienced at intervals of 12,000 years are divided by four Genesis stages: Tribal-Nomadic Genesis, Family-Human Genesis, Lesser Self-Genesis and Higher Self-Genesis, and Cosmos Genesis.

Tribal-Genesis	Family-Genesis	Self-Genesis	Cosmos-Genesis
2,500 years	2,500 years	Lower Higher 2,500 years	Lower Higher 2,500 years

When the Cosmos Archetypal Light reaches its cyclic fullness, the Spirit of God Eternal vitalizes His Mind-Energy overflowing to the Maya controlled world, and greater images of mind flow into the receptivity of those prepared. The total human race is repolarized and energized to learn directly of God and of His World, Purpose and Plan.

Always before the Cosmos Archetypal Light is ready to re-energize the soul-currents in the higher mind, men in the world undergo the latter-day great initiations which are called the Kali-Yuga time. Those evolved or Niscient are called to pass light to those knowing less or nescient. Christ Jesus is the pole or the balance-stay-way for those who serve in this great Maya day of fulfillment. Tribal-Genesis and Family-Genesis are now being readied to enter into the more direct Archetypal Light whereby a new human race will be produced. All will work to some degree in the new mind in Christ spoken of by Saint Paul.

In this earth, that which man sees by the eyes or hears by the ears or tastes by the tongue or smells by the breath or touches by the texture is of the ego as a Prototype. That which makes it possible to see, hear, taste, touch, smell is of the Soul and the Archetype. None of these things could be possible save that God has designed the Archetype which is in command of the Prototype.

Archetypal Grace enables a person to receive, telepathically, the greater ideas of God. Archetypal Grace has been earned through the ages by a person's repeatedly communing with the Eternal. When one has Archetypal Grace, he has become at one with the Archetypal Hum, and thus can speak of the Eternal with authority.

Prototypal Grace has been earned through the doing of selfless works in the world. Prototypal Grace enables a person to become at one with his soul and with the souls of all persons, and to achieve some act, work or creation of valor for mankind.

The Christ Spirit, speaking into the minds of those called to teach and to reveal, gives the greater Archetypal Knowledge to those having unblemished vehicles. The Christ Spirit is beginning to enter into the divine wave lengths of light and illumination of those who are willing and ready to serve, to heal, to teach and to reveal. Samadhi under Christ will enable pure and true initiates to articulate and activate what is experienced in the true Atma Light of Unconditioned or Absolute God.

The Cosmos Archetypal Knowledge occurs at 250,000-year intervals. This produces in man soul-knowing. The Higher Self comes forth to give God-awareness for all on the earth.

Christ Jesus opened to mankind in this planet Earth the Cosmos Archetypal Light when He ascended into the higher consciousness of the Father in Heaven, where They now work as one.

There are twelve prototypal images imaged by the Cosmos Mind of God within this eternity system. These twelve prototypes are maintained and sustained in their development through the zodiacal Hierarchy system. The twelve great constellations moving in proximity to this eternity system work directly within the Archetypal Light to shape, form and develop the twelve prototypes called *man* or *mankind*.

Man, as a prototype, is an intelligible being of sentient and consciousness life. His soul, which is a radiating Archetypal Center, is the most direct influence upon his intelligible thinking, feeling and acting.

God as Light works directly through the Archetypal Formulas or Blueprints shaping the mind of man. The Archetypal Blueprints move into the soul of man, where they are received and activated by his higher mind.

The Greater Archetypal System assures man of his ultimate destiny as a son of God.

The cyclic Moving Archetypes work directly with the planets of this earth and with the twelve zodiacal constellations in proximity to the earth. The Moving Archetypes work in cyclic compulsions and impulses lasting for 12,000-year intervals. Ten thousand years of a 12,000-year interval directly influence the prototypal man and mankind in his shaping and forming, and in his development through race and nations, through family and personality, and through self-identity or individuality.

Two thousand years of each 12,000-year Moving Archetypal Interval are interim years between each Moving-Archetypal Cycle. During the 2,000 interim years, man, while active in the physical plane, absorbs and reaps the benefits of his learning within the second or moving aspect of the Archetypal System.

The Destroying Archetypes work to destroy crystallization; they also work to keep the first two Archetypes in a state of constant circulation and action. The Destroying Archetypes enable men to be born and to die, and also prevent any form of crystallization to limit or to short-circuit the destined life-force of the reincarnation-tides and the vital Archetypal seasons and cycles.

As men progress in the earth, they are in a state of unceasing prototypal evolvement and change. Through the zodiacal balancing system, men reincarnate, that they may experience each and all of the twelve types within the wheel or mobile movements of zodiacal constellations.

Within the first aspects of the Greater Archetypes, men are assured a soul destiny. Through the Moving Archetypes, functioning as 12,000-year cycles, men repeatedly reincarnate, moving from life to life—and thus are offered the opportunity to achieve perfection within this solar and planetary system. Through the Destroying Archetypes, men are assured that their souls will move from cycle to cycle, life to life, until their destiny within an earth system is fulfilled within the Will of God.

The most profound initiation for man is the Prototypal Initiation. This takes place within the heart chakra. Through the heart initiation, one must enter into the one hundred and eight prototypal initiatory-gates.

In each life, one expresses a prototypal blueprint, called a *zodiacal Sun Sign*. His major work is to expand his consciousness through prototypal-association with other prototypes or embodied souls of the earth.

In each life, one is born into a family whose zodiacal prototypal blueprint must be incorporated and assimilated into his own nature. His affinity to, or his alienation from, any person is the expression of a need in him to know, to learn and to develop. Attraction to and repulsion by a personalized prototypal soul is the textbook material through which one learns in the physical-world school of life.

A perfect justice within the Love of God is assured in man through the Greater Archetypal system. The love of man through his own soul-powers, and through the use of the wisdom aspect of his higher mind, assures him of a benign and ultimate happy providence while living within the earth.

Aversion or hatred toward any prototypal person crossing one's path or associated with one in environment or circumstance produces karma. The Prototypal Initiation is unceasing from life to life. In each life, one seeks to incorporate each of the twelve prototypal-associated influences, and also master his own prototypal image. When one falls short of this, he is destined to continue to reincarnate under the heavier influences of the planets, the Sun and the Moon.

All men are linked together in the Prototypal and Archetypal plan within the Will of God. The soul of man acts as an Archetypal, superconscious, energized, zodiacal-reflected wheel. Man's prototypal development can be read and seen within the soul-light of the etheric horoscope. This can also be seen within the auric light by one having the power of zodiacal etheric clairvoyance.

Beyond the mathematical horoscope of mundane action, one must unite with the prototypal design in form and the Archetypal plan in soul. He does this through self-observation, self-examination, self-mastery and selfless love for every person mobilized and moving with him in his path of destined consciousness and light.

Man makes his fate and writes his karmic pattern upon the soul's record. His karma draws to him that person or prototype through which he must learn. Acceptance, love, appreciation, generosity and hospitality assure him of a relaxed and happy prototypal development.

As Jesus recognized His twelve, and became the soul-sun in their lives, so is each one destined to be a soul-sun centered within all twelve Prototypal and Archetypal projected forms.

God is the Sculptor; earth-man is the clay. The soul in man is the breath of life in the life of God. Over and over in this earth, man, through birth, repeats himself—improving, enlarging, and responding to the three great shaping and forming Archetypal compulsions.

Faith in God, in His Will, in His Plan, unites one with the Greater Archetypes, whereby the course for man, the Moon, the Planets, the Sun and the Stars is set. Faith in oneself as an imaged-vehicle in an unceasing state of progression toward individual perfection gives to each man or prototypal entity a happiness, a delight, a balance. When man is absent from belief in the Plan, he is but a puppet dangling from a self-will string propelled by ego and egotism.

The Destroying Archetypes, the third aspect for Archetypal expression, assure man that the little ego must die; that he is a progressive reincarnating prototype housing a soul. The Destroying Archetypes assure him of a degree of illumined consciousness—making him a creator with God.

Through the Destroying Archetypes, man is assured that every suffering and disappointment will have its ending. Through the Destroying Archetypes, man is assured that he is deathless, immortal. Through the Destroying Archetypes, one must accept that a friend will change and grow and develop, even as he himself is changing and growing and developing.

Looking through the eyes or vision of the Destroying Archetypes, one can be assured beyond the grave or death that he will be born again within the destined plan of the Greater Archetypes. Where he is born and how he is born will be determined by what he has done, what his attitude has been, how he has tried or not tried. All of these will determine what prototype he will express in coming lives and also what zodiacal prototypes will cross his path in coming lives. In each life, these prototypes will shape his outlook on life for good or ill.

Jesus was a perfected prototype. His life is the flawless example of the Archetypal System. As a Son of the Father, He dwelt within the Greater-Blueprint Cosmos-energies of Archetypal Light. He used this Archetypal Light in His higher mind to see the real in man and to see the destined intent of God in man.

The statements of Jesus are Archetypal Statements supported by the Spirit of Truth dwelling within the Greater Archetypes. Jesus came at a special time within a Moving Archetype; He chose to exist on the human level within a Moving Archetypal time. As a perfect prototype, He appeared among men; He understood that man was yet imperfect. He loved men in spite of their unknowing of themselves and of God.

Jesus—in placing Himself into the Moving Archetypal system, or limitation of Maya cyclic timing—accelerated the prototypal development of man, opening certain vital portals in the soul, mind and emotions. He prepared mankind for the Greater Archetypal Light of the Christ now beginning to enter the earth, whereby men will relate themselves to the Eternals and the Universal between souls.

Jesus knew that His time was limited on earth, for the clock recording time in a Moving Archetype limited Him, as other men are limited to the Maya-cycle units of measure of death and birth. In the Moving Archetype, man thinks through the consciousness of death and birth. Only when man unites with the Greater Archetypes is he aware of the Resurrection and the Eternals.

Jesus sealed men of the earth into the Greater Christed Son-of-God Archetypes when He was on the Cross. He could do this only when He was in the state of victory over death. Had he lived an ordinary life like any other, He could not have sealed men in, that they become Eternal and Universal in consciousness.

The Archetypal Seals are masses of vibrational energy-particles which are constantly being reinforced by the Archangels, the Cherubim and the Devas, who seal away the Greater Truths from the unprepared. Under Christ, the Hierarchy, the Archangels and the greater Devas open the Archetypal Seals when men are prepared.

Birth to realization,
Birth to knowledge,
Birth to the Invisible or the Inner is the great adventure provided by the Father in this Maya system.

SOLAR ARCHETYPE

17

THE CHRIST SPIRIT

The Christ Spirit is the Archetypal Presence permeating the Greater Archetypes.

A Solar Archetype

To understand and to respond to the Archetypal system of instruction is to be free. We who walk the Path today are more than mystical novices in light. As scientists we must look to Jesus, the Great Manifestor, who used the mighty Archetypal ideas and powers to manifest the Will and Mind of God to the world.

When we see Jesus and His miraculous powers within the frame of the Greater Archetypes, we can see that He truly did make the Word flesh in us through His mighty channeling of the Word of God. Since His coming, we are destined to increase our soul-expansion and our minds' light, becoming more aware in each age of the Word through Him. With a great age such as we are now experiencing, we are looking beyond the storm of chaos, determining to hear His voice above all others speaking for the Word of God.

All souls of the Earth, all creatures, and all sentient and non-sentient life are presently undergoing a synthesis receptivity from the great and mighty Archetypes or the Word. The Word is once again being made flesh or manifested into the soul, mind, forms, names and actions of man as never seen before in the Earth.

Through epochal and cyclical action, the Word during Great Intervals is sent into the world to receptive minds. Receptive souls respond; men rise to greater heights to create with their Creator—God. It is the destiny of all that one must produce from what he is (spirit). The Word or Spirit of God during a greater synthesis Archetypal time literally turns over the souls of men to compel them to do mighty and new, creative things.

During a synthesis Archetypal Time, the Solar or Sun Archetypes flow into the souls of men, uniting them with the Time-Spirit or the *Ancient of Days*. All who have followed the Light respond to this Solar Archetypal Light. The Christ, the Son of God, makes Himself known to those who are vessels for the Light.

The age now before us is literally a "time and times." In this great new age, lasting 2500 years, men will learn of and use the energies of the soul, mind and planets, thus becoming masters of Time and Space. From this will come forth another Golden Age to the Earth and to man.

When a Solar Archetype is quickened, this brings about the radical upsetting of decadent fixations. The Solar Archetype produces certain daring and creative inventive incentives until now dormant in the minds of men.

The Archetone within the Archetype sounds when the Archetype is ready for manifestation. The Spirit of Truth and the Holy Ghost stir and flood the upper planes of the Astral World or the First Heaven with ideas and pictures which fall into the mind of prophets, seers and advanced disciples; and this in turn produces the transcending changes in the outer life of the world.

THE CHRIST SPIRIT

The Christ Spirit is in command of the Greater Archetypes of the Mind.

All Archetypes are Christ-Mind Codes of the Divine. Through the expediency of need and through the versatility of one's innate logos, the initiate breaks open the codes within the Greater Archetypes. An open Archetypal Code is like the blasting of an atom bomb falling into receptive consciousness, exploding as Revelation.

When Revelation is on the Altar, the Mystic Christ is alive. If the teacher does not bear the Archetypal Seed, he is not a revelator. Every Archetypal-Seed Statement is a seed of truth containing power to open the revelatory aspect in the consciousness of the hearer.

Any teaching having a message must have a messenger, a messenger filled with Archetypal Light. Archetypal Light is Revealing Light, banishing ignorance obscuring the soul.

Hierarchy in other eternity systems works in combined light-sending to stimulate and give birth to new phases of Archetypal power in a new eternity system. Thus, Earth is a younger eternity system.

The coordinating Hierarchs or Overlords working with this eternity system work with Jesus, its selected World-Saviour, the Christ Spirit and the Father to bring to man a mental and spiritual identity only to be obtained in this Earth system.

Hierarchy works in ten labors. Ten per cent of man's gross and total income tithed back to the Lord and the Lord's work is the only means and way of man's reproducing the energy within the law of supply, of keeping open the freeing and wellspring of the ten Hierarchy Archetypal energies. The Substance or Image degree of the Greater Archetypes received from Hierarchy and the Father gives to man the supply of foods, grains, etc. When man returns his tenth through selfless channels to the Lord's work, he multiplies his seed or supply for the next harvest, thereby preventing famines, depressions, recessions and heavy taxation. In this, the unending thread of renewal in substance is sealed not away. Thus, when man tithes, he fulfills the first step and practice in discipleship and also frees the Melchizedek powers of future protection for substance and supply. (Genesis 14:18-20)

On all levels, men are striving directly and indirectly to extend their mental dimensions and perceptions. The mechanics of living in the human condition under the quickened Christed Archetypal Light now prepare to move men beyond the subjective yin ritualistically-dependent, habit-bound processes into a more direct perception of life and its meaning as creative and creating.

Cosmos ideals relate to the greater ideas overdwelling one eternity system. Cosmos Christ pictures relate to the Greater Archetypes. When one reads the cosmic ideals in the inner light, he sees but the reflected light. When one communes directly with the Greater Archetypes, he unites with the Spirit of Truth. *"Howbeit when he, the Spirit of truth, is come, he will guide you into all truth: for he shall not speak of himself; but whatsoever he shall hear, that shall he speak: and he will shew you things to come." (St. John 16:13)*

The Christ, until the coming of Jesus, worked with the Father, using the Archetypal rhythms to send the greater ideas to earth. The Christ centered His Light into the Invisible Sun surrounding the physical sun. When Jesus transcended the race-karma of the world, the Christ Spirit pierced the core of the earth. The Christ now works to absorb the World-Soul Atom so that all men may have the opportunity to receive His Light.

Our Father of this eternity works *with* the Christ. The Christ will take over the Father's work when the World-Soul Atom compulsions have been entirely absorbed by the

Christ Archetypal Impulse. This is what Jesus meant when He said, *"My Father worketh hitherto, and I work." (St. John 5:17)* When this occurs, men will be as sons of God or will have a creative consciousness similar to that of Hierarchy. *"Is it not written in your law, I said, Ye are gods?" (St. John 10:34)*

The Brahminic Archetypes which sustained the philosophies of the East are being withdrawn so that the Christ Spirit might come closer to all men of the earth. Thus men of the East are being forcibly separated and detached from their former philosophies. In the century ahead, there will be many radical experimentations, but the spiritual balance which seeks to be set in the earth will be sustained. Men of the East now die to the old and prepare for the new so that they may partake of a greater Light after communism has spent itself. Three hundred years will pass before the communistic powers in China will become less of tyranny and more of democracy.

The word *support* in the Rishi, the Christ, or Divine White-Line of teaching is called *Dharma*. Dharma-supports are established through merit-strength or grace earned on the physical plane by selfless works, sacrifice and giving; on the emotional plane, by processes in the use of love earned by union with the Love of God; on the physical plane, by use of energy-power extracted from Nature-energy; on the spiritual plane, by use of mind-strength and God-Realization, seeing with the Causal Mind God as the Creator.

The opening of the door to the Causal Mind is the ultimate goal for all on the Path. The Causal Mind is an unblemished mirror-vehicle for the Archetypal thoughts of the Christ-Spirit.

> *The Causal Mind is the non-contesting higher mind where peace reigns, where innocence omnisciently dwells.*

To be changed, to be born again through initiation to the mind, is to enter into the higher mind where the Archetypal Light of the Christ would give the new mind in Christ.

An Archetypal thought is a pure undefiled idea as yet unpolluted by any human concept. An Archetypal thought is a four-dimensional idea as yet unexposed to any form of competitive idea or comparison as to man's past thinking and acting. An Archetypal thought enters into a pure thinking process in which love, pure desire, and receptivity are wholly dependent upon God's revelation.

The advanced householder is preparing to become the initiate and, finally, the avatar. He must live in the threefold action of Initiations of the Dharma. In this, he balances the outer-world demands with the spiritualization of his intellect, and also he adheres to the Family-Atom Altar, to his research of Principle and Ethic, and to his knowing awareness in various stages of Illumination during and after Initiation. He is the threshold-teacher ready to take on the apprenticeship of direct Archetypal training in the Archetypal Dharma. He works in the world as a chela, self-acknowledged, as a responsible initiate in placement, whether work, home, family or discipleship. He becomes co-atom to his Teacher and, finally, to Jesus. Wherever he is, he represents the Jesus Ethic and lives it.

The hum of the *Niscience Archetype* is now sounding to the world and will bring a new spiritual impulse. The world was unprepared to receive the Niscience Archetype before the scientific age. The Niscience Archetype will unite men with the Jesus Ethic and will bring them closer to the Christ—thus producing in the scientific age a new theme of devotion and dedication.

There are many in the world who are influenced by the Niscience Archetype. Their hearts and souls are entirely dedicated to the Light of the Christ and to His humanity for the world. Such persons meet and gather together in the

Ministry of the Night, where they are instructed and directed by the illuminative Presences of Heaven. In the daytime action, those who have gained a spiritual dignity must show the way, having made a covenant in Heaven to work with the Equation of God.

Prophets have in a more or less degree always had their Angel of the Mystery yielding up to them through the prophetic sight the things to come. In the Niscience Archetypal era, more than prophecy is required, more than one Angel of the Mystery is necessary, that men of the earth be conditioned to the great age of terrestrial might. Only the initiates of Cosmos can reveal what is concealed from spurious sight. The macrocosm speaks through and into the Niscience Archetype, telling of the era of the balance between axis poles within the Universe. The Angels of the Mystery have loosed the seals of the Cosmos. Cosmos initiates now prepare to speak to the world.

> *And I saw a strong angel proclaiming with a loud voice, Who is worthy to open the book, and to loose the seals thereof?*
>
> —*Revelation 5:2*

THE ARCHANGELS

The Music of the Spheres playeth its work upon the earth. The voice of the angelic tones worketh with the nature spirits. The great Archangels arouse men to their spiritual oneness in God. The Light of the Christ shineth as a great eye into the eye of man—expanding his mind, his vision, his light. Man becometh more creative; more as a creator. He reflecteth upon the Image of God. He bringeth forth God's Light into earth; and the dark places of the earth become lighted places. Men find open places in all places.

What is set down so clearly in the Scrolls of the Archetypes is filtered down with the aid of the Great Archangel Beings. Men on earth receive the Archetypal Scrolls when they are ready to enter a new phase of evolvement.

The Archangels working with the Archetones keep man's faith alive. The Archangels use the great Tones in the Third Heaven to sustain man's belief in God. The Archangels working with other Angels keep certain layers of the heart, mind and soul awake to the spiritual verities of Heaven. Were it not for the Archangel-tides, men would fall into desolateness.

The Archangels cannot reach men who turn their beliefs toward physical exaltedness rather than spiritual exaltedness.

The Covenant of the Archangels is to work with God, the Father, the Hierarchs, the Christ and the Holy Ghost. The Archangels are the custodians of the Tones within the Greater Archetypes.

The Host of Archangels working with the Archetypes send forth the Archangels Raphael, Gabriel, Michael and Uriel during the four cosmic seasons of the year so that men will open their souls and their minds to the uplifting impulses of Heaven. Raphael works with the Vernal Equinox; Uriel, the Summer Solstice; Michael, the Autumnal Equinox; Gabriel, the Winter Solstice.

Gabriel works through the Judean matrix to bring science and the scientist. The sound of the horn of Gabriel in this age, or his Archetypal Tone in his angelic logos, is emerging as the telescope to unite man with the Galaxy Gate through which God reveals Himself as Spirit Omnipresence in all Eternities, Galaxies.

The Great Archangels Gabriel, Michael, Uriel and Raphael are the protective Guardian Angels of the prototypal symbols as sent forth from Hierarchy.

The Seraphim Guardian Angels and the personal Guardian Angels protect the initiatory symbols sent forth during meditation and contemplation. The Masters, the Illuminati and the Risen Dead work with the Seraphim Angels to prevent disclosure of knowledge out of timing.

The Archetypal Master Symbols in the Kingdom of God are protected by the Recording Angels, the Archetypal Guardian Angels, the Celestial Angels and the Archetypal Recording Angels.

The science of symbology is a sacred science given only to those who are at one with the Angels. Only the pure in heart may unravel the thread of symbology and earn the power to use symbols through ritual-formulas, and thus gain the power of manifestation and de-manifestation.

Symbols proceeding from the Greater Archetypes are spiritual vibrationless symbols and are beyond defining as to their degree of light.

The Archangel Metatron, the Solar Archangel watching over the Earth, seals in the Light of the Christ. During the darkest time of the Earth, the Winter Solstice, he begins to seal in the great Archetypal Light of the Christ.

The Archangels go before all men in new beginnings. They assist in the birth and death of planets; in the birth and death of all seed life. They assist in the birth and death of man. The Archangels work with the birth and death of races and nations. They aid the Angels of Birth and Propagation. The Archangels work with all things which have a seed in earth. They work with the human seed to be born; and they work on the sacred-atom seed of the heart which enables man to die.

The Archangels of Saturn aid men to cross the vista of death—and death becomes a peaceful sleep and repose. He who dies receives his recordings through his personal Guardian Angel, his Recording Angel, and through the Lords of Karma; and receives his peace through the great love of the Archangel power.

Saturn on men's death giveth to them the record of their memories and the power to rise above purgatory where they reflect on the Life Archetype and its Tone. He who knoweth the key to Saturn hath reached the Tone in the Archetypal World and can in vision and observing watch the working of the Tone.

Illness in the Earth, caused by the temporary separation from the Archetype, recordeth itself as discordance in the emotional or astral world. He who hath obstructed his ego-tone through that of the weak findeth within his body sickness and discordance. He who is utterly free hath placed Saturn as his keynote in the highest place of his being and hath exalted it out of the many days of knowing the Father. For he who hath known the Father in many lives and been at one with Him hath Saturn exalted.

THE MEDALLION OF THE MASTERS

The Masters are working with the Son of Man through the Greater Archetypes.

Jesus is the Door through which one enters into the Kingdom. Jesus is Cosmos Dimensional. To reach the Kingdom, one moves through Jesus, connecting himself with the Omnipresence tributary units of His Cosmos Light. The tributary units are the Saints, the Angels, the Bodhisattvas and the Masters.

The Light Current, the Life-Stream Current and the Sound Current become one when one has made union with the composite Masters who now work with the true initiates aligned with the Greater Archetypes. The Masters come to keep men in the *Word* or the Light Current, the Life Current and the Sound Current.

The great Masters in their days of ripeness have known four great Ages in the Archetypes of Man. In the Holy Scriptures, the epochs of the life in earth are spoken of as *the days.* The epoch in the life of man is spoken of as *the generations.*

The Archetypal Masters are presently overdwelling the light, life and the souls of those who are called to teach, to heal and to reveal what the Greater Archetypes are saying to men of the Earth.

The Masters have the power to use a unique telepathy between soul and soul—a telepathy which no human tie can separate or disorganize. The telepathy between the disciple and the disciple, and the disciple and the Master's Light, is never severed or cut away once the bond has been proven in other lives.

When one is in the Master's Light, he is co-atom to the Master and receives an uninterrupted flow of telepathic thought from the Master. When one is in the love of the Master, he is in a state of grace and receives the telepathy of the Master through his emotions and intuitions.

The disciple in Light moves beyond vibration and sees in a higher degree of Light. He sees that all things contain an ultimate and a completion after the plan of the pure in creation. The things before his gaze stand transparent and transfigured. He beholds all things transformed in the light of the Master's Light. He learns that Light carries with it the music and the tone-sound of the Archetype within the Master's Light. He also learns that he who has touched the Master's Light has tapped the Intelligible Soul. The Intelligible Soul is the Rational Tone or Archetypal Tone within the Archetype, determining the beginning, the survival and the completion of each thing under the influence and design of the Archetype.

The Masters' Medallion is of the great molten sea which spreads the Archetypes as they work with the Will of God. He who stands in the Masters' Medallion touches the Archetype as Tone; spreads the Tone as rhythm; spreads the rhythm as sound; spreads the sound as cycle; spreads the cycle as the time of the days. He who stands in the Masters' Medallion becomes as a timeless one and sees the All from the beginning. He penetrates Nature's realms, the Kingdom of God; and all things in the earth become as one consciousness moving upward to God.

In the Medallion of the Masters, the disciple's solar plexus is protected and the astral tumults are sealed away. The spiritual images within the brain are reflected directly above each temple, even as the horns of Moses, which Moses used to bring forth the flame from the bush. He who carries the etheric horns carries the power of the Neptune's staff in the earth, the power over the elements, the power of discerning Nature and the Nature Spirits, the power to live upon the ether of Nature and to subsist on less food; he carries the power to draw the vital energies from the air, the power to explore the Archetypes within Nature, and to correlate the ideas of man with the Archetypes of the Spiritual Worlds.

In the Solstices and the Equinoxes, the cleansing winds come to man; they bring a renewed electricity, giving man new beginnings and new magnetisms, that he may work and walk within the Medallion of the Masters. He who has stood in the cleansing wind and has heard the voice of the sylphs, the music of the Angels, the soul-tones in the Spiritual Worlds, and the Archetones of the Greater Archetypes has heard the music of the Kingdom of God. The cleansing winds affect the atmosphere, that man might be free of the elemental forces of the dark, and that he might be cleansed of ancestral and tribal karma.

The *Venerable One* is a Bodhisattva Master-Saint. He is centered within the hub or center of the Great-Soul Medallion of the White Line of Masters.

The one called G.Q., the Little Brother, is spiritually close to the Eastern, yogi initiate and also to initiates in the Western world. G.Q. equalizes the flow of ancient Archetypal ideas. He is the last of the true Masters, even though considered as the least of these. The Western initiate, being unable to assimilate the yoga practices and formulas, is enabled, through the Master G.Q., to retain the basic truths within the ancient Archetypes of the Vedas, the Sutras, the Upanishads—for example, the law of birth and rebirth or re-embodiment. The laws of eternity justice, or karma, are made clear to the Western initiate through the work of the Master G.Q. He works with the disciple of the new age to retain the old, and, thereby, to gain the whole in Christ.

The tone of each living thing in earth correspondeth to a tone within the Archetypal Worlds. If one multiplied billions of times the finest tones of a musical instrument, he would find the tones in the Archetype from which form moves, extends, changes, blends, and taketh its part in man's station.

The inner ear within the Higher Worlds recordeth one tone. The inner ear holdeth the tone and never loseth it. It retaineth it unto all ages. It is this tone which worketh with the recording of the soul; it bringeth man to his own Archetype from which many lives have been sent forth, and from which many prototypes have been projected. Thousands of experiences, griefs, joys, creations, failures—all have proceeded from the tone which still resteth in the center of the ear.

Man in his progression in thought, silently withdrawing himself into the Archetypal Hum or Inner Tone, placeth himself in the memory of this Tone and draweth it forth. He calleth it silence. The Master calleth it Tone.

THE MAHA SUNN OR GREAT VOID

The Archetypal mind-seeds are infinite God-Mind molecules active at aeonic intervals and spacings. Their vibrationless tones and light function in the Great Void or matrix of the Cosmos Deep.

Man, as a prototype as to type or form, must from aeon to aeon mark, trace and claim his potential.

When Elohim, the Dual-Gods or Hierarchs, sound their Tones and send all of their Rays together with the Father into the higher mind of the earth, under command of the Christ, the earth comes to a receptivity-hush. The prototypal samskara-seeds of partial knowing through sense action must be dissolved and erased. The war between the dark and light of the soul and mind begins, for the souls of the earth must prepare for a virginal action and knowing of the mind.

When the Void containing the Godhead Archetypes overflows into the world, all conceptual falsity is accelerated and erased. Men appear in evil and in sin; their works of the dark are dredged forth. Their souls are ruthlessly exposed to frictions beyond sense, beyond astral, beyond lower mind. This then is the beginning of the New Day, a New Day of Hierarchy to create for Hierarchy and with Hierarchy and to gain the Archetypal goals impressed and compressed into the mental matrix of the mind, the soul.

In preparation for the Archetypal Light, the Lords of Forces—the Devas embodied within the atmosphere of the gunas—make the karmas of all souls rajasic, heated, passional, fiery, that birth may be experienced to a new mind. This mind is the new mind in Christ promised by the Apostle, Saint Paul.

Those penetrating the Archetypal Light have experienced the Great Void through Christ-consciousness experience. He, the Son of God, is that One now lifting the veil.

As God is Creator, man is creator. As all beginnings must begin with a Void, so must man begin with a Void. In the Vedas this is called the *Maha Sunn*. No one of earth experience can bring to birth in the outer consciousness the firsthand knowledge of the Archetypal Light save he has first crossed over the darkest Void, which is the darkened side of the Sun, reflected and mirrored into the Purusha (Over-Soul) Void.

In crossing over the Purusha Void into the Void of the Greater Archetypes or Maha Sunn, one enters into the fathomless as related to conceptual mind (lower mind)—returning from the Void of Maha Sunn into Maha voids of earth phenomena or relativity.

Through Christ, the distinguishable wisdom is retained. He who is Christ Jesus is that One who holds the thread to lead the initiate over and through the terrifying, glorious splendor to God. He is that One who gives to the initiate the Archetypal thread-sustaining to remember, to transmit, to reveal.

Samskara means impressioned seeds of karma grown and developed through knowing in part, through wrong use of mind, emotions and actions. The samskara karmic seeds dwell in the innermost interior of the subconscious mind. In great New-Day beginnings, their activity is magnified so that they may come under the Archetypal erasements. During these times, the evil-clinging and craving must be transposed so that the lesser ego can become a supporting rather than an obstructing vehicle for the Archetypal Light.

> *In the Archetypal aeonic birth, the soul of man, woman and child will receive from the solar rays uniting with the Solar Devas, Hierarchs, Archangels. They will cross over the Maha Sunn beyond the lower and dense Maya friction voids into the Spaceless and Timeless terrain of the Eternal.*

MANTRA POWER

MANTRA—Speaking a Name of God or one of His aspects. Spoken repetitiously, produces purification of the mind, giving mind-steadiness, making the mind a pure vessel for Truth. Repetitive speaking of one's own mantra is called JAPA and produces Self-Realization.

MANTRAM—A mantra sentence directly touching the need or concern during initiatory processes in the physical, astral and etheric planes.

Pure and true sutra, mantra, bija, and mantramic function is made possible when one makes alignment with the Great Archetypal formulas provided by the Will of God within His Son, the Christ.

In every Yuga or Age, there is a fresh or totally new Archetypal downpouring of seed-formulas within the Word of God. In the present age of Kali-Yuga, or dark time of spiritual ignorance, the great primordial Masters are sending new and pure tones to the chelas and advanced disciples of the earth who are responsible for the guardianship of the Dharma. The new sound of the Word, or of the Archetypes within the Word of God through the Christ, falls upon the receptivity of those who are being initiated to become the bearers of the Dharma or instructors of the spiritual life.

In the next Yuga-Age, called *the Golden Age*, men will enter into a more direct awareness of their spiritual and true natures.

The sutra, mantra, bija, mantram and prayer are Tonal or Archetone passwords to the great Word-Sound holding together this eternity system and its contributing aspects of consciousness within the Will of God.

A prayer, when for self, contains no sutra meaning, no mantra seed-tone, no bija instruction, no mantramic formula correlating to the greater Archetypal Tone and Archetypes. Such prayers are futile and sterile, mere lip-service without life, future or vision.

The Lord's Prayer as given by Jesus in this time of spiritual darkness, or the Kali-Yuga Day, contains the sutra, mantra, bija and mantramic functioning power.

All Archetypal techniques of purity in sutra, mantra,

mantram and prayer—when given by one who has crossed over the abyss of darkness of the Maha Sunn and has returned to distribute to men the *Danda* (Law), the *Mercy* (Agape), and the *Dharma* (the Instruction)—must be accepted and mastered by mankind to divert destruction in the world.

One should speak the Lord's Prayer with uttermost awareness of the concealed crescendo-power contained therein. This Prayer, when used with the lips rather than with the subjective-heart awareness, is as a dry seed of no-sound sounding in the wilderness of non-comprehension and ignorance.

Speak and live. Awaken the vita-nature of the sound within each word. Become the *Word*, and thus become a precious gem-tone for God.

The *New Sound* and the old sound of sutra, mantra, bija and mantram are now fusing to produce a New Way, a New Era in which the Christ will be known as not any other is known.

Joseph is the Master Carpenter and Jesus is the Lord of Prospering. Joseph is also the watchman over the Archetypal Blueprints supporting the New Covenants or Dharma. The 7 is his number. He initiates his proteges through the formulas of 7.

> *When I have mastered the mantramic words, the Tone of the Archetype shall make my mouth a sweet well of Heaven.*

THE HIGHER UNCONSCIOUS

> *Wherever there is an eternity system or a solar system, the Greater Purusha or Higher Unconscious, which is the Mind of God, controls the rhythmic and cyclic processes of life solar, of life planetary, of life lunar and of life conscient.*

The *Greater Unconscious* is the Spirit energy-field of the Unmanifest where resides the Archetypes. Christ, as Overlord of the Greater Archetypes, as Presence, is the Archetypal Omnipresence within the Greater Archetypal System of the Purusha Higher Unconscious.

The Archetypes and the Archetones presently are functioning in the hierarchy nature of the Higher Self and the Buddhi nature of the higher mind. This is the realm of the *Higher Unconscious,* or that place of stilled quiet which sounds the Audible Hum of the Cosmos within the Eternal Spirit of God.

Where the spiritual world meets the gravity world, there is a spherical realm of pure mind and soul atmospheres. With extended senses and soul-faculties, man living in the gravity world can explore the Archetypal Realms of bliss. He can make himself an experiencer of this bliss-vicinity by purification of the chakra light in his nature.

The first phase of initiation into the Higher Unconscious bliss-realm is perilous. In this, one must encounter the struggle aspect of the Undersoul that he may still or quieten the frenzied energies of the ego, the intellect and the lower mind and will.

All who seek spiritual experience are destined to retain something of gravity in their articulating what is seen and revealed. In the search for spiritual experience, one seeks to pass on what he has seen, heard and experienced; it is his work to pass on to others that they may also know and experience. One works that he may move over the gravity obstructions, that he may be united directly with God, and thus learn of the spiritual mysteries at first hand.

Archetypal Knowledge is available to those who have moved beyond the gross gravity energies in their lower chakras. Beginning at the heart chakra, one relates himself to the Higher Unconscious through Meditation, Prayer, Contemplation and the Holy Practices. In the Higher Unconscious or Buddhi light, the Archetypes flow directly and rhythmically into the souls who seek to know and to experience God as Eternal Spirit.

All persons on the plane of earth through the laws of karma and reincarnation are not equal or the same karmically. All souls in the Archetypal Kingdom of the Greater Unconscious are equal in soul and soul capacity.

The great Laws stemming from the Unmanifest use selectivity for that one who is worthy, pure of heart and hands, white as snow in Ahimsa. This selectivity is divine and cannot be utilized as an ego-vehicle for self-claiming or fame. The selective laws of Archetypal Power are by Eternal Law in the Spirit of the Eternal *bound to happen.*

Once one has developed the Ahimsa heart, which has been tried through the aeons as to faultless serving and giving to God, he is ready to experience the Greater Unconscious.

When one has some remaining pleasure for gain in his emotional nature—and has other merit states of worth in his nature—the Higher Unconscious Archetypal Light moves upon him and his karma as devastation. Intense suffering is the result.

There is a difference between the Greater Unconscious and the Collective Unconscious. The Collective Unconscious is made up of the entitized humanistic-will experience; it is a *memory-body* recording experimental creativity in the earth. The Greater Unconscious is the Purusha or the field upon which God radiates His Spirit as *Knowing* for man.

Subconscious mind draws upon the past which is present as active soul-memory in the Collective Unconscious. Archetypal Presence represents the *Spirit of things to come.* All revelation, all prophecy on the level of the Greater Unconscious contain the future Blueprint signals and portent for things to come in earth and in Heaven.

He who has earned an Ahimsa heart moves over the bridge of the Collective Unconscious containing formats which men have proven and disproven.

The seer as a prophet crosses the bridge of the lower subconscious, which relates to gene-memory and the reincarnation soul-processes of the past. Having mastered these, he walks beyond the abyss into the Bridge of Light, where he makes union with his hierarchy nature and becomes an imager-creator for God.

Everyone who has been *opened* to the Higher Unconscious where dwells the Archetypal Light has the power of *Passing*. The Holy Spirit makes such persons into holy seers and revelators.

Where reverence is, God can abide. One must first have *Altar-reverence* to experience *Altar-awareness;* from this, he has Soul-Transmitting power. Then comes the Pure Logos, repeating the Archetypal Word.

Word-formulas and sound-formulas, given telepathically from the Higher Worlds, will accompany the new Archetypes of the new age. New vowels, new words, new objective forms in speech are to be designed within these word-formulas. Men shall create, more and more, through speaking and sounding. Speech as power will be used by those who understand the telepathic word-formulas given from the Higher Worlds.

> *The Archetypal Worlds are actual soul-mind seed pastures, where we go in and out clothed in our invisible or seamless garment.*

Solomon's life is the example of "so far and no farther" —for some persons in the world may attain much wisdom, but they act emotionally through their tribal-genesis feelings. Such persons are unable to have access to the Celestial Archetypes, or the power "to go in and out."

The Ancient of Days, as the Preserver of Time and the units of measure, is in command of all etheric encasements in this earth system. He is in control of timing as to the length of the functioning of a Race, the length of the years in one's life, the length of an Archetype's sounding and sending, and the length of mankind's mortal days.

Determination to live draws upon units of measure in the Greater Archetypes which in the God-nature are Timeless as to days and years.

Sin, which presently is seemingly so impenetrable by the Light, is to be exposed by the Light of the downpouring of the Archetypal Truths. The blind-accepting ignorance of man will be transformed, opening doors of astonishment to the tamasic and the rajasic mind.

The rajasic heats which have overwhelmed the mystic body in the church and in the teachings must now return to a most pure, holy bliss consciousness within the Logic rationality with the Archetypal Omnipresence.

The present brain of man is controlled by the Moon. The advanced ego has both the Sun and the Moon influencing his brain. Only through Archetypal Instruction does an Initiate open the Gleaming Brain which is fed by the Sun.

The pineal gland in man acts as the matrix for the Sun energy. The pituitary gland acts as the matrix for the Moon and the Lunar energies.

The three steps of psychic power are initiatory steps. One is in the primary grades as a lower lunar psychic; in the secondary grades, as a higher lunar psychic; and as a solar initiate, one has moved out of the algebra of mystical metaphysics into the higher physics of Archetypal Dimensions.

In Pro-Genesis, men will be manifestors, having earned the Melchizedek powers—the power to increase and the power to decrease. Their thoughts will be scintillating manifestors for God. Pro-Genesis souls will work directly with Archetypes to be used when this earth has begun again a new eternity day.

We are now heirs to the Greater Archetypes facing a new human-race cycle.

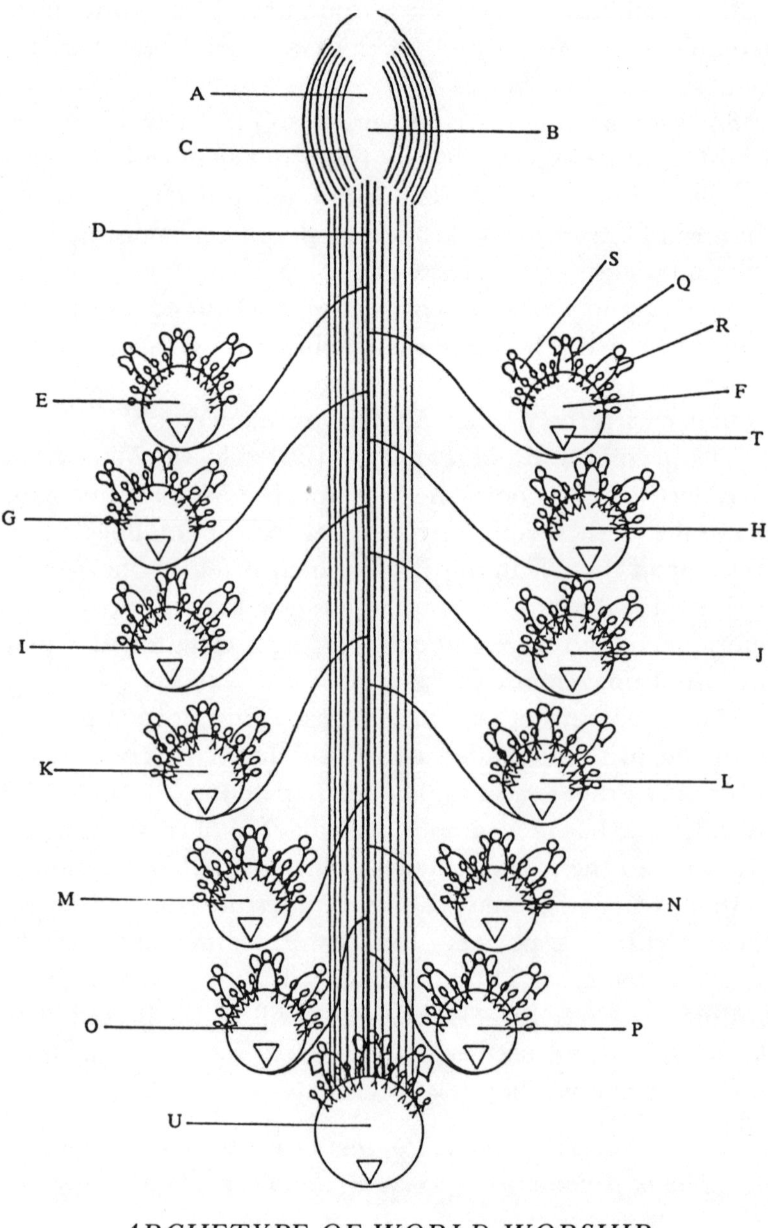

ARCHETYPE OF WORLD WORSHIP

18

ARCHETYPE OF WORLD WORSHIP

Jesus was called "an high priest after the order of Melchisedec" (Hebrews 5:10) because He was a manifestor and a de-manifestor. He is the high priest over the Archetype of Worship.

Worship

SUNDAY WORSHIP PRAYER

Let me heed this day—
to be nigh unto prayer.
Let me be in the company
*of prayer*full *souls.*
Let me find the Comforter on my knees.
Let me be filled with the
Glory of our Mighty God.
Amen.

When a person has no desire to worship on the Day of Worship, which is Sunday, he has shut himself away from his Destiny and is resting in the lap of karmic fate. One who has the grace of longing for worship has the assurance of union with God in this lifetime.

Persons who do not understand the Law of Karma worship at the altar of phenomena. Such persons become the victim of the lusts of others and their own lusts within.

He who worships *the god of appearance* will walk the path of losing everything that he may find the way of the Real. What he calls the real will rot in his hands, including his mind faculties. He will live in unceasing decay until the face of the Real will finally be shown to his eye, for he is a slave to Maya, to the illusion that makes a *static imbecility*, seeing *everything* as remaining the same.

Creeds are separative, calculating specialization of individual choice in worship. In the ending days or Kali-Yuga times, creeds go sour—turning on themselves, producing dissensions and abnegation; altars become political rather than spiritual. The clique system in creed produces agnosticism and opportunism for self-inflation, and irritation on the sensitive in awakened souls in a congregation.

The Holy Spirit is ever at work to eradicate soiling upon the Holy Altar. Reforms come; old liturgy-dependencies disappear. Materialistic ego-interpretations of God speak from the holy places. Jesus' Name is rarely spoken. The Jesus Path is feared as it leads to Intentional Suffering.

Creed-bound religions, above all enforcing separation, suffer extermination, for God-Increase as Holy Spirit spews out the damnation edicts separating soul from soul, man from man.

The caliber of a worship is revealed in the testimony.

It is not one's religion, it is what he *believes* that gives him eternality. One's beliefs are his religion. The Worship Principle resides in man first as belief; and, second, as faith.

The credo of a worship body should be: To minister rather than to be ministered to.

Worship of God in the Family Atom in the life comes first.

Melchizedek is the most high priest over the Sacraments

of Worship. His place is in the Second Heaven. The Comforter is in the worship.

The Sacrament is the redeemer and virtue-revealer. Bread by the woman and Wine by the man produce the whole polarity Wine-and-Bread offering, whereby those who take the Sacrament are gauged by the Whole Unconscious or the Collective Unconscious, which is the vehicle for the Holy Spirit.

> *Each day drain off, purify the senses through meditation and contemplation. Each day study a portion of Scripture. Each day give at least one gift of your compassion. Each day serve one afflicted person.*

THE ALTAR

> *Due to a preponderance of Antichrist activity in the world, religion faces within the next 2,500 years the greatest trial of its history.*

Religiosity is a zeal to know, to recollect and experience God. Redemption is recollection. The stirring of the Holy Spirit quickens religiosity. One is redeemed, and thereby recovers his remembrance of God.

Unholy nescience is atheism experienced throughout many lives. The hardened heart against the Altar-place where the Omnipresence is found results in a return to sadism, hedonism.

Men generous with prayer and nongenerous with giving-works are like worms eating the dust from decaying scriptures.

"And everyone that hath forsaken houses, or brethren, or sisters, or father, or mother, or wife, or children, or lands, for my name's sake shall receive an hundredfold, and shall inherit everlasting life." (St. Matthew 19:29) A true

minister is looking for his hundredfold where his Altar will prosper, that an hundredfold will become a thriving handful of God, whereby the Providence of God will be known in its superabundance of spiritual vitalities.

The *Jnana* is experienced through reading, writing and spontaneity creation in the use of the mind. The *Bhakti* is experienced through service to the Altar of God, service to the human race, and worship with a devoted heart. These two must be incorporated in the Western Initiate's householder mind and heart to fulfill their soul-requirement in this era and time and in the magnificent eras before them.

When Jnana and Bhakti are united, this is the whole man in Christ, each given its own proportion, that the Holy Spirit may fulfill its nature of holiness in perfect equilibrium.

Ahimsa gives harmlessness so that the Bhakti intuition can work.

Some people have a *wild* heart. They go in many directions in their choice of love—and dissipate their love. They are going in so many directions that they are yet to experience the Ahimsa and Bhakti love. The Ahimsa and Bhakti love is a patient love, knowing how to use the energies of self-discipline for the selfless way.

The reasoning mind of the Bhakti heart makes few mistakes, for that which works through them knows only one direction: God first in the spirit of love, and all else is mercy.

> *God is a Spirit: and they that worship him must worship him in spirit and in truth.*
>
> —*St. John 4:24*

THE ARCHETYPE OF WORLD WORSHIP

One of the major Archetypes centered in the Third Heaven is the Archetype of World Worship. This Worship Archetype is under the command of the Christ Spirit, the

Holy Ghost, the Elohim or Hierarchy, and the Father. In the Archetypal Kingdom, or Third Heaven, this worship impulse correlates to the twelve prototypal images of the zodiacal Hierarchy. When the Holy Ghost sounds the Archetone or Eternal Hum within the Archetype of Worship, this is received and expanded in the centered apex of the Third Heaven, where dwell the twelve Apostles of Christ Jesus.

The Third-Heaven worship impulses or mobile worship energies overdwell the sacramental etheric blueprints in the Second Heaven. These impulses fall into the saintly prayer cloisters of the First Heaven, and finally, into the conscience of man. The conscience of man compels him to seek God, to worship and serve Him. Man proceeds in his worship through conscience; through saintly and angelic helps of the First Heaven; through Seraphim sacramental initiation in the Second Heaven; and through the Archetypal Christ Light in the Third Heaven.

In the Second Heaven, where all sacrament ritual begins, the priesthood is sifted by Melchisedec, that it may become a pure priesthood. There are twelve priestly prototypes; each pure priest correlates to one of the Disciples of Jesus. Each priest must eventually earn the apostolic power and be sanctified and ordained under the Christ. Such priests keep alive the Word of God.

In the Third Heaven, each of the twelve Disciple Prototypes correlates to a Hierarch, a mighty Judgment Angel, a Recording Angel and a Guardian Angel. When a priest of an earth church is ordained under Christ, he is reborn through the Holy Ghost; he receives the apostolic authority; and he then becomes a mediator and a revelator for the Third Heaven.

The Christ assigns to each worship impulse a Hierarch, a Judgment Angel, a Recording Angel and a Guardian Angel. A congregation of worshipers is made up of initiatory units of twelve. In every unit of twelve worshipers there will be eleven faithful, and one who will make himself a reflector of Judas. Through these Judas members, the satanic power will work unceasingly to give unrest to the other members of the congregation, that they may be spurred to research their faith, their hearts, their purpose.

The Archetype of World Worship is a blueprint of all religious, mystical, occult, metaphysical, philosophical, esoteric and spiritual compulsions as expressed and experienced in the world of man. The first ethical technique of teaching is to instill into the ones being instructed charitableness toward the worship of others, that they may sacredly and reverently acknowledge and give to all men the equal right to worship as to their conscience, timing and placement.

The following letters (A, B, etc.) and their accompanying explanatory notes correlate to the letters on the Chart on page 364. The purpose of this Chart is to inspire charitableness and understanding toward all forms of worship and religious seeking.

A The Will of God works invisibly through the Central Core or Life Vortice of the Archetype.

B The Life Vortice is the work of the Father and Hierarchy, and is the vehicle for the Images. The Images are sealed into the Eternal Sustaining Atoms. Our Father, working with the Images, uses the Will Principle and the Life Principle. *"Let us make man in our image, after our likeness." (Genesis 1:26)*

C The 12 Archetypal Flames around the Life Vortice keep alive and sustain the Images within the Eternal Sustaining Atoms. The Christ is the Lord of *all* Archetypes and uses the Archetypal Flames to channel Light to the earth. *"Let there be Light." (Genesis 1:3) "I am the Light of the world." (St. John 8:12)* Each Archetypal Flame makes a vehicle for a great Principle.

There are 12 Archetypal Flames and 12 great Principles in each Archetype. Through the 12 Principles, pure Ethics are communicable to man and sustained in earth.

D The Archetypal Thread or Arche-thread is the correlating thread between man and God, and is the means through which Jesus, using the Love Principle, makes entry into each vortice-pool of worship. *"I am the true vine." (St. John 15:1)*

E to P The 13 vortice-pools of worship, projected from the "vine" or Arche-thread, represent the various ways men seek to find God. In each vortice-pool of worship, men of similar nature, by the law of attraction, seek to outpicture personal and impersonal attributes of God. One may identify the type of prototypes within each vortice-pool by the type of Hierarchy overdwelling the vortice-pool. There are three Greater Beings of Mediation overdwelling each of the thirteen vortice-pools: (1) a great Hierarch Being, (2) a Guardian Angel and (3) a Recording Angel. There is also a sub-being or *Dweller* in each vortice pool. The three Higher Beings and the Dweller fulfill a quatrad action, thereby giving FORM to the worship.

Q A Great Being or Hierarch correlates to each vortice-pool of worship. The prototypes in each particular vortice-pool are aligned to the "likeness" of the Hierarch. This determines the type of worship or expression within each vortice-pool. The Hierarch sustains the prototypal image which men in the vortice-pool are seeking to perfect through worship.

R There is a protective Guardian Angel over each vortice-pool of worship. This Guardian Angel is in direct alignment to the *needs* in men's worship, and overdirects and times *when* and *how* men receive the Archetypal ideas, principles and rituals. The principles are received as *ethics* by those more highly evolved and as *moral laws* by those yet to reach higher evolvement.

S There is a Recording Angel over each vortice-pool of worship. This Recording Angel records and enscrolls onto the World Medallion of Grace the good works manifested in each vortice-pool.

T The Dweller of the vortice-pool of worship is created out of the sins of the mass in each vortice-pool. The Dweller is an entity-body built out of the prejudices, hates, bigotries and hypocrisies yet to be resolved in each pool of worship. The Dweller retains unto itself the down-note or malicious and evil works existing in all religions and worships from whence come persecutions. Through the Dweller's action, men, in certain cycles, feel compelled to persecute one another in the name of religion or worship. *"It must needs be that offences come; but woe to that man by whom the offence cometh." (St. Matthew 18:7)*

In periods of dormancy when men leave one another to their worships, the Dweller in each vortice-pool is kept in abeyance and balance by the Guardian Angel. When a particular vortice-pool no longer serves the purpose as an expression of pure worship and alignment, the Guardian Angel withdraws; and the Dweller, using the Destroying Principle through dissension and strife, dissolves the activity expression—and the religion or worship ceases to be as of its kind. The Dweller also works to periodically cleanse groups or mass worships which become over-crystallized. Such Dwellers build a barrier impasse around each vortice-pool through ceremonial magics built by repetition in ritual. An individual desiring to evolve into another phase of expression in worship is challenged by the Dweller of the former vortice-pool; and on relinquishing his association, he must contend with this Dweller. The Guardian Angel of the vortice-pool of his new action in worship will protect him if he proves himself truly sincere and seeking toward higher fulfillment.

U The Christ-Spirit overdwells the 13th vortice-pool. The 13th vortice-pool is activated only during exceptional periods of great evolvement, that is, when mankind as a whole works to make another spiral in Light. Until the coming of Jesus, the 13th vortice-pool was only partially manifested (e.g., the Essenes).

Religious impulses are predominant in the first 12 vortice-pools. Spiritual impulses are predominant in the 13th vortice-pool. Through the 13th vortice-pool, the religious is kept balanced with the spiritual. In the 13th vortice-pool, advanced disciples are drawn together due to their intermediary powers and works of Light. Those who work with the 13th vortice-pool seek to increase the spiritual impulse in all worships. The worship-ethic and activities will influence all other teachings in the world, inclusive of education, science, religion, philosophy, metaphysics. Such persons should be aware that they are the direct recipients of a new and vital action, and that they should not compare the instruction with the instruction received in past vortice-pools, but they should strive to qualify themselves to become purified channels and serve God.

The 13 Vortice-Pools of Worship in the World Today

Instruction: To expand your horizon of charitableness, it is recommended that, as you read the lettered paragraph describing each vortice-pool (E, etc.), you refer to the correlating letter on the Chart *Archetype of World-Worship*, page 364 and also refer to the Contemplative Exercise at the end of this chapter.

E Emphasis: reform; conversion; evangelism. Purpose: those expressing the reform or evangelistical emphasis in worship are working to cleanse the inherited ancestral or Tribal-Genesis pictures received through the emotions. This works through moral law and conscience. Down note: bigotry.

F Emphasis: duty. Pertains to Human-Genesis association or that which holds men together through blood. Purpose: to fulfill the moral law in the Ten Commandments. Down note: possessivism; cunning will expressed through Tribal-Genesis instinct.

G Emphasis: men worshipping as their fathers worshipped. From this stem the patriarch religions. Out of such worships come men who become pioneers in new communities: Purpose: obedience. Down note: inclusiveness.

H In this vortice-pool, men are austere and express literal religions. Emphasis: virtue. Purpose: to sustain good works. Down note: opposition.

I Emphasis: the worship of form and materiality. Men in this relation to worship revere goodness in man as their God. Purpose: to achieve integrity. Down note: aggression.

J Religions that build their worships on ritual and fraternities that sustain certain truths through procedure in ritual. Emphasis (religions): reverence. Emphasis (fraternities): honor and responsibility. Purpose: ritual makes it possible to sustain certain truths that they be not lost to the world. Down note: crystallization and isolation, and the dangers of ceremonial magic through the helps of elementals used in ritual.

K Scientific; psychological; yoga. Emphasis: mental will and mentalism. Purpose: self-control. Down note: atheistic.

L Emphasis: religious mystic, retreatist. Purpose: purification of the personalized emotions. Down note: the psychical.

M Emphasis: prayer and healing. Men in this vorticepool are unrealistic and impractical, yet some of these have the sustaining grace of reverence as the means through which healing is manifested. Purpose: faith. Down note: blind faith and belief in an effortless escape from karma.

N Emphasis: organizing of the thought world. A trial vortice unique to our time, in which man may be initiated into the lesser mysteries or formula disciplines in thought. The purpose is to devitalize and detach the former ceremonial magic impulse sealed into the lesser etheric body through many ages of religious and formal worships. In our time there are many off-shoots in this trial vortice, all classified under the name of metaphysics. This metaphysical expression may be compared to the time before Jesus in Greece, Rome and Egypt when men, through their self-wills, worshipped the many gods rather than One and thus scattered their energies in worship. This also carries the danger of the belief of self-omnipotence rather than self-reliance and that which relies on God, the One. Down note: irreverence.

O The initiatory schools, visible and invisible: occult; philosophical; esoteric. Emphasis: (occult) power. Purpose: (occult) personalized initiations to command the emotional will, and invisible initiations to rise above the lesser regions and energies of the Astral World and to make penetrableness to the higher planes of the Astral World. Down note: (occult) the use of black magic and inverted powers through exploitation or curiosity.

Emphasis: (philosophical) impersonality. Purpose: (philosophical) observation in the physical world and perception into both higher and lower abstract worlds. Down note: (philosophical) intolerant and critical mind, and intellectual pride. Emphasis: (esoteric) ethics. Purpose: (esoteric) to produce a continuity between the physical, etheric, emotional and mental worlds; to aid in the overcoming of the tumults in karma in the lesser regions and planes of the Astral World; and, through various initiatory processes, to qualify through pure intent and pure motive for higher discipleship powers. Down note: (esoteric) the loss of spiritual power and danger from impersonation.

This vortice-pool is one of the last preparatory stages of evolvement before entering the 13th vortice-pool and thus carries heavy penalties of karma. Such initiatory experiences may extend over many lives before pure reverence, pure dedication may be reached.

P Emphasis: searching. Those in this vacuum-like vortice-pool are the "homeless ones" or those without a spiritual home. Purpose: to isolate themselves from the many human interpretations revealed through formal worships, and to seek to find the Real and recover their sacred relating to God, that they may begin their serving in Light. This vortice-pool and phase of their evolvement carries with it a recapitulation of everything experienced in former lives as of religion, metaphysics, philosophy, occultism, and the heavy burdens of karma. Through such karma the answers will be found and eventually they will enter the next or 13th vortice-pool where, with knowledge and experience in equal degree to reverence and ethics, they will blend with the harmony of pure disciple relatings or that "prize of the high calling of God" (Philippians 3:14) and the standing in the Light of the Christ.

Down note: Detaching themselves from the effort and grace as pictured in their souls' records and therefore becoming agitators wherever they are placed in their karma.

U Emphasis: Fulfilling the dedicated discipleship practices and qualifying (1) to serve in perfected alignment to the Higher Self and to the power of mediation in the Spiritual Worlds and also in the telepathic association with co-disciples in earth, and (2) to become a healing minister in both day and night serving. Purpose: The dedicated disciples in the 13th vortice-pool are the most highly evolved and spiritual beings in earth. Their combined alignment to one another in atom association makes it possible for them to work, visibly and invisibly, in direct alignment to the Love of the Jesus One, the Archetypal Flames of the Christ, and the Purifying Principle of the Father. The works of Light consummated by these sacred disciples balance and sustain the love, light and potential spiritual impulses in all of the vortice-pools of worship. Through the 13th vortice-pool and the use of direct Fiat action, such dedicated disciples—wherever they are in the world—blend together, and as living arteries of spiritual light translate and transubstantiate the new and virginal spiritual impulses onto mankind. Personal karma, family karma and world karma blend as one discipline that such truly devoted and dedicated ones, whose labors are wholly selfless and non-personalized, may work in world serving. Down note: failing to sustain the alignment to the 13th vortice-pool.

CONTEMPLATIVE EXERCISE

Instruction: To sustain the note of Charitableness throughout your spiritual endeavors of the future, it is recommended that you contemplate a different vortice-pool each day. For example: "E" the first day, "F" the second day, etc. During the remainder of the month, use this same procedure and recapitulate the contemplation of each vortice-pool in its sequence as part of the whole Archetype of World Worship. The following seven steps may be used in each day's contemplation. There should be no tension or critical analysis, as this is not an Exercise for intellectual comparison but to incorporate, through contemplation, the living, rhythmic activity in worship as expressed by men in the world today. This Contemplative Exercise may be used any time of the day suitable to your convenience.

1. Contemplate the type of action in each vortice-pool as a living and unending stream of activity.

2. Contemplate each vortice-pool as part of a natural order and sequence as to the stages of man's evolvement through worship in its varied expressions.

3. Contemplate the Over-Being, the Guardian Angel and the Recording Angel over each vortice-pool so that you may incorporate this beneficent activity as a reality.

4. In your contemplative thought process, emphasize the Jesus' Arche-thread or "vine" keeping alive the love-impulse in each vortice-pool, thereby making men, in spite of their seeming separateness, one in God.

5. Contemplate the great Principles and the Light of the Christ.

6. Contemplate the Archetypal Flames which sustain the life in each vortice-pool and the work of the Father.

7. And then contemplate the whole, that is, the vortice-pools, the overdwelling Beings, the Jesus One, the Archetypal Flames, and the Archetype as part of a perfected Plan created and activated tnrough the Will of God.

The above Contemplative Exercise is for a one-month period only. However, if you feel the need to review this Exercise in the future, it is recommended that you adhere to the one-month rhythm as given in the above Instruction.

19

ANGELS AND ARCHANGELS

One who enters the Archetypal Path cuts the cord of the pleasure-principle expectations and takes on the golden cable of oneness with God.

Animal Species Archetypes

And the Lord said unto Noah, Come thou and all thy house into the ark; for thee have I seen righteous before me in this generation. Of every clean beast thou shalt take to thee by sevens, the male and his female: and of beasts that are not clean by two, the male and his female. Of fowls also of the air by sevens, the male and the female; to keep seed alive upon the face of all the earth. For yet seven days, and I will cause it to rain upon the earth forty days and forty nights; and every living substance that I have made will I destroy from off the face of the earth. And Noah did according unto all that the Lord commanded him.

—*Genesis 7:1–5*

The story of Noah has caused much controversy and confusion because its allegorical meaning has been degraded into fable. There have been numerous cataclysms in the earth and inspired survivals overdirected by angelic helps, particularly in the earlier days of humanity when men were more aware of their etheric relating to God. The physical life of Noah correlated to a great cataclysm and the withdrawal of an inverted type of humanity into a twilight state. Noah, one of the Elect or advanced beings of his time, was made the custodian of a new animal life-wave and a new humanity life-wave. The "ark" described in the Bible relates to the spiritual "*arc*hetype." Noah, being the pure, prototypal representative of a new life-wave of humanity and a coming rise in Genesis, sustained the thread of life-survival for animal and man of the earth. In all new beginnings or new life-waves of the earth, the Animal Species Archetypes and the Man Prototypal and Generation Archetypes work with a cleansing and extracting action. In the ancient times, these cleansings were always accompanied by great tidal waves or cataclysms.

Every 10,000 years, masses of laggard people are withdrawn into a twilight state and certain of the obsolete species of the animal kingdom are also withdrawn. Prehistoric animals, such as the dinosaur, etc., were withdrawn in previous cataclysmic times.

When a withdrawn species of the animal kingdom is seen by the disciple in the dream world or in meditation, he is observing and researching an ancient period. By the animal or creature seen, he may determine the period he is observing and researching.

Each Animal Species Archetype has its Species Guardian Angels. For example, the feline species has the Species Guardian Angels to control and time the mating seasons and to protect the animal in birth and death. When an animal dies, it is taken into the Species Guardian Angels' etheric anesthesia and, thus, does not suffer pain in death as compared to man. In the case of one animal destroying another animal, the captured creature is anesthetized by the Species Guardian Angels and the death pain is nullified. When man kills or destroys an animal, he has violated the Commandment, "Thou shalt not kill." The animal suffers pain because of this violation and is not protected by the Species Guardian Angels. When man kills for lust, for the sport of killing or domesticates an animal, thereby imposing his karma upon the animal, the animal is isolated from the help of the Species Guardian Angels. The Species Guardian Angels are not the same as the Species Elemental Guardians.

Animal Species Archetypes reside in the second region of the fourth astral plane, where they are overdwelt by the Species Guardian Angels. The Animal Species Guardian Angels work directly with the Cherubim Angels, producing all animal symbols in meditation and dreams. Man's moods and temperament, which so closely relate to the astral regions and planes, correlate directly to many animal correspondences. In this, man may be seen to express attributes in his character which are similar to particular animals. These similar correspondences between animal and man may be seen especially in the habits of man, in that a man may be as selfish as a hog, stubborn as a donkey, cunning as a fox, timid as a hare.

NATIONAL ARCHETYPES

In the Inner World Governings, each nation has a Blueprint or Archetype which directs the nation toward a certain maturity and fulfillment. Should the Archetype be offended, the result is that certain aspects of the Archetype are withdrawn; and men of such a nation must re-establish their works to the higher impulse working within the National Archetype.

Men first began in the earth as tribes. From tribes developed races. From races developed nations. In the present period of man's evolvement, his soul and his physical world are engrossed in a national scene.

Each nation's particular Blueprint or Archetype correlates to the polarity of the nation as to latitude and longitude of the earth.

Each National Archetype has three Angels: (1) the Guardian Angel who guards and protects the Archetype of the nation; (2) the Recording Angel who records the actions within the nation, seeks to sustain the ethic within the nation, and stirs the conscience of the nation; (3) the Archangel of the nation's Archetype who keeps the nation open to the Greater Archetypes in the Higher Worlds.

When a nation is involved in a war, the conscience of the nation is being cleansed. When men refuse to respond to the higher ethic within a nation, the Archangel presiding over the Archetype withdraws the life-spirit from the nation —and the nation falls into the hands of the enemy.

Each nation has a core-identity stemming from its Archetypal polarity-point. Ultimately, nations will respond to the Heavenly Archetypal pattern of Ethic in government, of laws to support the rights and the individuality of the growing spirit within each one and each living thing.

The Recording Angels working with the Archetype of Nations are the spiritual chronologists of nations. They record the purpose and intent of a nation, and record the valorous and heroic events in a nation. They retain the historical records of nations so that patriotism may be sustained.

Each nation has a heartbeat which is centered in the magnetic pole, reflecting its Archetype and holding together the boundaries of the nation.

Every nation is a polarity-center for the arising of a new mankind. There will be troubles and tumults until men unite themselves with the spiritual governings of the Archetypal World.

The Archetype over each nation seeks to produce individual initiatives, whereby the incentive to be individualized through creation produces vital-life in a people.

Nations must be healed. Nations must free the Archetypal initiatives and incentives for *all* of their people.

When the tribes had been established, Hierarchy and our Father began to sound higher tones of generation into the Archetypes overdirecting the human spirit. Each man began to be the chief of his own house. In this period, the tribal encasement began gradually to change to Family-Atom encasement. Man began to learn the rugged lesson of family survival, rather than the survival of tribes. No longer dependent upon the chief in a tribe, he had to rely upon his own resources for survival. He began to learn that he must choose his own mate and protect her for the sake of their children.

Race and racial placement fulfill the Will of God. One should revere his placement in race and the placement of others in their races; and he should remember at all times that there is an invisible law governing all races.

One should understand that nations live and die, even as men live and die; that nations may reach heights in their destinies; and that nations may err, even as men—but one should look to the Archetype of a Nation and see its intent for the many in national placement.

> *Turkey standeth in the present chaos as the polarity in the turning point of men and life on Earth. She standeth as the key magnetic pole, working counterclockwise, for remnant Race Archetypes, and is closest to the rhythms of the overrule of the Race-Angels.*

One having the power to unite with the Archetypes of the Continents can see the old decaying etheric-molds of former civilizations. These continent-molds act as the cradles for the birth of new civilizations. In this age, these etheric molds have ceased to give virility to the life-exploring aspects within the lower chakras of man. Through the use of the higher chakras, the seer sees another thrust coming to birth in each continent—a thrust of new and vital awakening, pushing aside the decay, that the vital essences of birth may have their play within the Plan and Will of God.

Governments which have swept men into corruption and have brought crises to men in earth; systems within the systems of government which have beguiled themselves after the desires of a few, ignoring the necessities of the many; religions which have formed of themselves cliques, constructed strongholds, fortresses—these sound a discordant note in Heaven, offending their Archetypes. The tones of their Archetypes now begin their ceasing and, in this ceasing, their action is being withdrawn from the earth.

Each governmental city of a nation has an Archetype. This Archetype influences all surrounding cities. Those who have soul-greatness in governmental cities work with the Archetype of the city. The disciple who lives in or visits a governmental city communes with the Archetype of the government and learns of the ethic supporting the nation.

The city of Washington carries an androgynous polarity. Those who strive for power in government for selfish purposes corrupt themselves and their soul-lights' future karma. The Archetype in America seals itself off from the evil waves which dash upon it from time to time out of men's cunning minds.

The Archetype of America overdwelling the city of Washington is set with the Great Seal. The Great Seal disturbs not itself from the time of its beginning unto 1300 years. When it has reached this time, all forces may come out and contest it. The Seal remains, and those who break the laws set within the Seal break themselves.

The karma of the United States of America has not yet its inscription in the recorded Scrolls of Light affecting governments. Not having fulfilled her Archetype as yet, she remains as a cradle of chaos.

The *Great Mountain* is the earth's master-mountain, which is seated in a certain latitude and longitude of magnetic sensitivity. This mountain is a heavenly radar for the Spiritual Worlds. Until the year 1950, the Himalayas were the center of the earth's spiritual polarity. Due to the withdrawal of the Brahminic Archetypes from the Asiatic world, the present site of the Great Mountain is Mount Whitney, in the United States of America.

People must return to the Archetypal Morality. In all nations, those who sit in authority must be at one with the authority of the people.

MANTRAM FOR THE HEALING OF A NATION

> *May the Angels of the Archetype of this Nation inspire each one who dedicates himself to fulfill the pure intent within this national community of peoples so that all men of this nation may be blessed of God and sacredly prosper.*

GENERATION ARCHETYPES

> *There is an Archetypal Record of Creation. There also are Virginal Archetypes of unborn Nations and Continents; Archetypes for the Humanities of the earth; Archetypes of the Animal Kingdom; and Archetypes of the Plant Kingdom.*

The reach of the World-Soul Atom encases the Moon, the planets and the Sun. Thus, the Earth becomes the wife of the Sun; the planets, children of the Sun; the Moon, the discarded placenta of the Earth.

The World-Soul Atom extends its influence into the planets and the Sun. This makes it possible for the eternity system to move in its orbital and correlated life.

The Cosmos Atom in the Sun never ceases to control the World-Soul Atom centered in the core of the Earth. The Cosmos Atom, working with the World-Soul Atom, transmits to the Earth the Archetypal and Generation Prototypal impulses of Hierarchy and the Father. When this occurs, the Sun has found its rhythm; the Earth, its polarity; and the planets become bodies of mediation and unison suitable for stimulating the primary mineral and atmospheric substances and elements for life on Earth. The Species Angels overdwelling the Animal Kingdom stir the Generation Archetypes of the animals to take form in the Earth. The Terrestrial Angels, working with the Flora Angels, revitalize the Flora Archetypes to quicken the plants and foliage of the Earth. The Generation Archetypes sound the tone of life in man; the etheric form of man is then clothed in a physical body; the glands and arteries are simultaneously ripened through the pressures of gravity.

Singing of great music in dramatic form, such as opera, dissolves certain karmic molds of human conduct. Folk music and folk dancing unite one with the Generation Archetypes and the memory ancestral streams expressed in families and locales.

There is a kinship between all living things. God uses the Greater Archetypal Fiats or Archetone tone-commands or His Word vibration to send forth in varying degrees of tone the species blueprints for flowers and plants.

The Soul's Pulsation is the centered vortex or higher-mind outlet for the Higher Self and the higher etheric body. The heart is the mind outlet for the emotional body and emotions. The great lunar brain centered in the solar plexus is the mind outlet for the lesser etheric body. The sacral and sexual center is the mind outlet for the physical body. Each mind outlet contains a certain intelligence. All outlets seek to work with, to utilize, and to protect the vitality of the lesser etheric body.

The sexual-mind outlet seeks to reproduce progenitor children of the body. The Archetypes of Generation and the Propagation Angels work with the Moon or lunar tides to stimulate the generative or sexual processes of the body. Average man at present thinks through the generative sexual aspects first because he is geared to start all processes through generative compulsions. He secondly thinks, intuits and feels through *his karma*; as his karma is registered and sealed into the impressionable, etheric, karmic imprint patterns psychically photographed upon the lesser etheric body's impressionable mold. He is compelled through his karma to attract to himself the needed lessons and situations held over from past lives and in the present life.

Man next thinks through the mind outlet in his heart. Through heart lessons of love of a selfless nature, thinking of the heart within the heart produces a selflessly motivated life.

Man thinking through the higher mind at one with the soul's pulsation is a free initiate in the higher echelons of Heaven. His thinking is a creator profoundly rational within his higher mind; he is the designer within the Great Design.

Therefore, all men, when faced with the greater initiations, must recapitulate in the true and natural procession as ordained in the pattern of this eternity life. First, sex is weighed as to attitude and motive. Secondly, karma must be balanced and erased as to the physical aspects and former utilization of the mind; all indulgent aspects in character, emotions and mind must come under correction. Thirdly, the love nature must be adjusted to enlarged horizons beyond the love of self. Lastly, the mind is exposed to the credulity aspects and phases of thinking. The thoughts as an amoeba chain must be searched to find the weakest link in the chain, thereby dissolving and washing away all circumscribed limitations in thought.

All presently in the Archetypal Path must become devotees true and whole in devotion to live as an Archetypal statement and representative in the world.

SCIENCE AND THE ANGELS

Men of science in the earth hold the key to world brotherhood. Men of Light and spirituality, and the spiritual Beings within the Light of God, hold the key to spirituality and the key to life. Between these two, the balance standeth. Science seeketh to know the ultimate in its physical counterpart of the Spiritual Worlds; and the spiritual knoweth the Real in the Archetypes within God.

He who blazeth a spiritual trail is prompted by the compulsion of the Divine Archetypes. He relieth upon the pure law of reasoning within the direct Light in God.

When men in Earth are desolate, the White Brothers of the higher order come forth to minister unto them with the aid of the Angelic world, that their intellectual souls be lightened that the Truth enter and the stabilizing patterns in the Archetypes place themselves as formulas in men's consciousness.

When men are caught into the cross currents of dying to the old and being born to the new, there is great struggle; and the necessity for physical survival produces an extreme pressure upon the unprepared. Thus, the Angels come forth to aid the hapless in periods of Archetypal Chaos, or that is, when the old Archetypes are being withdrawn that the new might come forth.

Everything, before it appears to the eye on the physical

plane, must pass through a birth matrix, and this birth matrix is an energy-void through which the Archetypal work is done by the Cherubim. The Cherubim especially work with man's desires. The *cupid cherubs* are a branch of the Cherubim. They work with the planet Venus and the emotional body.

When the time is ripe for men to receive the Archetypal secrets relative to the form of each thing in the Earth, the Terrestrial Angels unite with the Celestial Angels who dwell in the Archetypal Kingdom. The Terrestrial Angels or Gravity Angels of the Earth work with the Cherubim Industry Angels of the fourth plane of the Astral World when man is ready to activate or give form to the Archetypes overdirecting mechanized action, such as science, machinery and mechanics.

> *Science but reflecteth the Archetypes of the Higher Worlds, and hath not the heart of the living life within it. Science is a temporal means through which men experience and experiment. It carrieth not with it that love, that ease, that peace, but it addeth to the noise and the clamor, upon which the first astral plane doth set up a cloud, a sea of repulsion.*
>
> *Science now holdeth the world at war or at peace. It is in science that the spiritual life and light must come; for men holding the key secrets and having direct access to the reflected archetypes endanger the earth. They carry the most sacred treasure which, if loosed onto profane hands, could violate that which God hath set into the earth.*

The Greater Archetypes in the Kingdom of God remain quiescent until men are ready to activate them into works for the world good. All ideas which men usually believe to come from their own inspiration are, in reality, received from the reflected archetypes residing in the fourth plane of the Astral World. The reflected archetypes are protected by the Cherubim Angels. Should one have had access to the reflected archetypes in former lives and have offended the ethic as to the use of such knowledge, he would be inclined to exploit this knowledge in the present time. The result would be self-destruction and destruction to those who place their fate in his hands.

In the Atlantean record there are records not revealed to men pertaining to the age in which men will recover or rediscover laws and ultimates not yet used in the lives of men. That which is recapitulated as of Atlantis is of a higher spiral as to science, but of a crude spiral as to perfection. The creative substance used by Atlanteans was of a malleable material perfected directly under the Masterbuilding Archetypes. Atlantean men, so close to the direct contact with the tone of creation, achieved a certain perfection. The latter men of Atlantis profaned the law in the use of the Archetypes. These forfeited the right to use the skills and experience them over many Epochs to come for men.

The perfected craftsman of the human spirit will be a combination of builder, shepherd and scientist. He shall be so rooted within the vine of the Son of Man, or Jesus, that he will incite a cosmos metamorphosis in the souls of others; he will be an eternal man inhabiting an eternity. Having ripened and matured through the ages, he will express the immortal tone of love. He will embody a versatile composite of ethics on all levels of human association. His mentality will draw upon the higher ethos of humanity, and he will give formulas to soften the harsher aspects in human nature. He will have unobstructed access to the compendium of the Spiritual Archetypes or greater ideas stemming from God.

The return of the Archetypal Knower-impulse preserves the advanced souls so that they might propagate and produce pure progeny who enter the world aware of stewardship within the Preservation Laws. Every ethical disciple must seek to spiritualize his intellect, that he might comprehend Omnipresence or the Spirit Eternal governing and determining the Way of Life.

Awareness of the Father is the last break-through the initiate must experience before he may truly serve the Father. The Father doing all things through him, the initiate wills to restore life; he wills to open the Archetypes; he wills to manifest, to de-manifest—all through and by the Father's Will in him.

In the process of the progressions to come in the future years, men shall name the Archetypes by many names, and shall acknowledge their existence by many conditions—but men shall not call them the Divine Archetypes. The scientific world shall give a name to the Archetypal system, for science shall recognize that there is an Intelligible Will bringing all things to perfection by degrees.

> *Within the next three hundred years, Greater Cosmos will possess men, even as Lesser Cosmos has possessed them until now. When men come under the influence of Greater-Cosmos ideas for an extended period of time, a new age beginneth in the world, and the souls of men respond to the Light of God.*

20

GLOSSARY

ABSTRACT WORLDS—The higher abstract world consists of the alignment of thought through light and tone. The lower abstract world consists of lesser degrees of light and sound.

ADI-BUDDHA—The Buddha-Being overdwelling all Buddhas. Adi-Buddha is also called Amitabha-Buddha. He is an overdwelling Presence stemming from the First Source of God, inspiring Buddhahood. The title Buddha means *awareness*. Such awareness is of the transcendent state which is beyond mental phenomenal, beyond the ego mind and also the lower myth-concepts.

AHIMSA—Harmlessness. *Ahimsa* is mastered in the four chakras below the throat. This is gained by transenergizing lust to purity, hate to love, greed to giving, separateness to oneness.

AKASIA—The pure, vitalized, supernal spiritual life-substance used by consciousness. The akasic light is soul-light. The color of akasia is indigo blue and is seen often in visions during meditation. Ether is the coagulating life substance. Prana is the energizing life substance in ether. Akasia is a light *consciousness* animating substance. Akasia substance is used in the highest form of spiritual telepathy. The more highly evolved one is, the more akasia he has in his mental and soul-light. One breathes in pranic energy to renew life force. One meditates to increase akasic light in his mind.

AKASIC RECORDS—Records of former lives.

ANDROGYNOUS—Not to be interpreted as the physical hermaphrodite. Androgynous is of a state of life consciousness which will blend the body, the soul-light and the World of God into man's physiology by transforming the positives and negatives within the physical expression into the whole man of Christ. Androgynity in this age is improbable. In the coming age men will approach a divine rapport between the lower self and Higher Self, and become androgynous.

ANGELS—Angels come in companies, legions and armies to succor the needs of man. The Angels are at home in earth, in the Astral World, in the Kingdom of Heaven and the Kingdom of God. They minister to Nature and to man. They are angelic ambassadors for Jesus, the Christ, our Father and Hierarchy. Knowledge of the Angels and initiation into their world of action may only be received by the disciple through perfect alignment with the Jesus One.

The *Personal Angels* watching over man's personal life work with suggestible angelic induction, not as man works with suggestible mind. The Angels of the body, emotions, mind and destiny cease not in their suggesting reminders to man, reminding man that he is a cognitive creature learning and earning within a phenomenal world.

The *Celestial Angels* work through initiates and adepts to reveal the inner wisdom-truths of God.

The *Cherubim Angels* initiate man into the mathematical processes in the use of mind plasmas or akasia.

Union with the Cherubim makes one a protege of the cosmic order system within the Universals.

The *Seraphim Angels* initiate man into techniques of light, whereby he comes in contact with the Archetypal ideas as given by the Christ Mind. The Seraphim Angels lead the mind of man into the six dimensions or the door to the Third Heaven, where he is then initiated by the Archangels into the Archetypal Light and the Archetones under command of the Holy Ghost.

ANCIENT OF DAYS (see Sanat Kumara)

ANTAHKARANA—The bridge between the lower and the higher mind, taking one upward to the *other shore*, or the bliss-state in consciousness.

ARCHANGELS—Work to establish continents in the earth and to lead the masses into certain polarities in the world. They work with the Seasons, Solstices and Equinoxes. Four of the Great Archangels—Gabriel, Michael, Uriel and Raphael—are the protective Guardian Angels of the prototypal symbols as sent forth from Hierarchy. The Archangels make it possible for souls to flow into other eternity systems.

Gabriel is the Archangel to the Southern Gate, the Gate of Enlightenment. The sound of the horn of Gabriel in this age, or his Archetypal Tone in his angelic logos, is emerging as the telescope to unite man with the Galaxy Gate through which God reveals Himself as Spirit Omnipresence in all Eternities, Galaxies. He who comes under Gabriel as an Initiate moves with the Archetone of Gabriel through which his soul-gifts will magnify and purify the human condition in the human race. The Immaculate Conception in Mary, officiated and manifested by Gabriel and the Holy Spirit, signifies the birth of the Higher Self-Genesis in the human race. The human race is reaching toward the birth of the Omniscient Cell in the Gleaming Brain. When this occurs in the Earth, scientific and spiritual creators will be cognizant of Galaxy-functioning as affecting life in this Solar System.

Michael is the Archangel to the Western Gate, the Gate of the Law. He is the Lord of the Archangels. During the solemn Autumnal Equinox, the Archangel Michael works with the great Judgment Angels, that the conscience of men in the earth may be reached; the souls of men come under judgment. In the Autumnal Equinox, men must meet their errors and acknowledge them.

Raphael is the Archangel to the Eastern Gate, the Gate of Spiritual birth. On the last Saturday before the Vernal Equinox, Raphael begins his holy work for the birth of the Christ Mind through the powers of the Resurrected Jesus.

Uriel, the Archangel to the Northern Gate, works to give death to the ego. In the Summer Solstice, Uriel begins his work with the human race. The will, when it is in the Will of God, is as a scintillating snowflake highly energized with God-energies. These energies are impressionable through consciousness. This will is directed through the Northern Gate through which Uriel comes. Uriel is the initiator of the higher mind used through the will in God.

ARCHETONE—The Word under command of the Holy Spirit used in conjunction with all Great Archetypes. When the Archetone sounds, the Archetype comes alive with whatever compulsion is ripe or ready to be manifested. Thus an Archetype remains dormant until the Archetone sounds. The Holy Ghost working with the Will of God sounds the Archetone, that the great ideas or new impulse for man may enter the world or fall into the mind of man in timing to the Will of God.

ARCHETYPAL CORD—The four life cords are the *umbilical cord*, the *silver cord*, the *Kundalini Sushumna cord* and, the highest, the *Archetypal Cord*. The Archetypal Cord is an actual vibrational cord through which the Holy Spirit recharges the soul of man. The Sound Current or Audible Hum vibrates within the Archetypal Cord. This Holy-Spirit sound containing the Holy Name of God, when free, sustains illumination and revelation. Also, when the Archetypal Cord is functioning in conjunction with the Sushumna, the inner Kingly cord of Kundalini and the silver cord are as one; one is in command of the transcendent processes within Buddhi or the Informing Principle, whereby he receives increasing guidance and spiritual vitalities on all consciousness planes functioning as soul, mind, emotions and body.

ARCHETYPE—Original or divine blueprint for the mind and life processes in the earth. The Greater Archetypes work first as the Unmanifest or as the Word yet to be made flesh. These spiritual, creative, divine Archetypal compulsions move in tides into the world through the minds of men. In each great Archetype is an Archetone. The Holy Spirit sounds within the Archetone when the divine Word is ready to enter into the mind and life of man.

ARCHETYPES, GREATER—Concern the greater lifewaves directing and changing reincarnation tidal flow; also, determine the timing of the appearance of new ideas in the receptive minds of the Elect or Adepts in preparation for new eras in time and evolvement.

ASTRAL—Star or planetary reflections. The astral is a unique, unceasing, fermenting, mirroring and moving action producing in man a state of emotion, mood and inductive feeling and thinking.

ASTRAL BODY—Emotional body.

ASTRAL CORE—An ovoid shaped envelope of astral fire kept in balance by nadi points or sound vortices located along the spinal system. The nadis act as absorption points receiving the sound and vibratory currents from the planets. The astral core represents the emotional body. The planetary fire playing upon the spinal system is distributed and generated by the assistance of the glands and the nadi points. Thus, the nervous system is protected in the etheric and physical bodies from being the recipient of direct vibrational force playing upon the spinal canal. When the astral core is unprotected, the glandular system goes out of balance, and the emphasis is an overcharge of psychic energy in the etheric body and in the lesser mind. One protects himself from the unruly charge of the astral-core fire through mantrams, contemplation, meditation, prayer and selfless service.

ASTRAL WORLD—The Astral World has two parts, seven planes and forty-nine regions. The *lower* Astral World relates to man's subconscious mind in death and in life. The *higher* Astral World is the First Heaven, where dwell the Buddha, the Bodhisattvas and the higher Devas or Shining Ones. The Astral World is supported by the magnetic belt around the earth. The grotesque level of the Astral World reflects the sin-body of the earth. The four lower planes of the Astral World are the recipients of the lower vibrations and energies of the planetary light. This sub-planetary energy produces a chimera mirage-like effect upon the emotions of man in life and in death, producing glamor and mesmeric effects upon the lower mind and senses of man. In the Bible, the Astral World is called "the serpent." Man commands the Astral World by the use of his higher will and higher mind. Before one can enter into spiritual illumination, he must undergo the trials of the subjective astral states within his subconscious mind. Saint Paul refers to the chimera action of this astralized emotional play when he speaks of "seeing through a glass darkly." (1 Corinthians 13:12) The Astral World is also spoken of in the Book of Revelation, Chapters 4 and 6, as a "sea of glass." When one masters the astral consciousness, the sea of glass is "no more," as it is dissolved, that one may reach the Second Heaven.

ATMA—A Divine Particle or a portion of God; the Eternal Sustaining Atom containing the Higher Self. It is also called *Atman*, meaning the Real or True Self; the source of all wisdom and beauty in life.

ATOM ASSOCIATION—Atom association is the means by which karma is hastened and resolved, as the ego is placed in an environment where through karma-sentience he intuits and selects the associations to which his atoms either correlate or are antagonistic.

AURA—The energy field around the body; the magnetic and electric field encompassing and encircling the body of any living thing. In the average man the aura is ovoid-like and extends approximately six feet. Herein reside the portrayed feelings and thought pictures sent from man's experiencing as of feeling and thinking. The Aura of the Master consists of a body of luminosity and ether which is the means of his remaining within the vibration of the earth and also the Spiritual Worlds. The disciple touches the Master's Aura through meditation, and in this manner receives the telepathy dwelling in the body of the Master's Auric Light.

BARDO—The three lower astral planes experienced after death.

BHAKTI—Devotion. Heart Initiate.

BIJA—Each chakra has a central atom (seed), called in Sanskrit *Bija*.

BINDU—A master secondary chakra. The *Mouth-of-God Chakra* situated at the base of the skull; centered in the medulla oblongata.

BIOTRON—Biological pranic-energy. Dualistic energy flow affecting the polarities.

BODHISATTVA—A Saint with transcendental powers supported by selfless serving. The destiny of the Bodhisattvas is to elevate men from misery to happiness. Their vow is to continue to reincarnate until all men of the earth make union with God. Bodhisattvas have lived in the physical world as holy and sacred men. They now work in the Nirvana planes or 6th and 7th planes of the Astral World. Their labor is to still the astral tumults shading the soul-powers of man. Among these are the future saviours who will enter the earth every 2,500 years so that man may refresh the spiritual impulse in the world. Such saviours are not to be compared with Jesus, the Messiah and Saviour of the world.

BRAHMAN—Unconditioned God; Absolute God.

BROTHERS IN WHITE (See White Brothers)

BUDDHA—Awareness. The one known as Buddha, who lived approximately 500 years before Jesus, now dwells in the First Heaven, called *Nirvana* in the Eastern teachings. Buddha is in command of the First Heaven. He acts as the Emotional Body of the Earth. The mastery of the kinetic, biological, astral, psychical and psychological emotional forces is under the crown of Buddha. Buddha gave to mankind the Eightfold Path by which one could unite consciously with the Cosmic Mind. Jesus gave to all men of the earth the opening of the twelve chakra-gates of the soul whereby man could consciously unite with

God as Cosmos Being. Buddha gave to man Mind-Realization and Self-Realization. Jesus gave to man God-Realization. Through God-Realization, man achieves Being rather than ego. The Path of Buddha and the Path of Jesus are both essential Path-ways to the Esse Dharma, or the Real. Both may be simultaneously realizable when one begins to move beyond the secular religious vestments representing Buddha and Jesus. Jesus is the Door; Buddha, the Path to the Door.

BUDDHI—The Informing Principle. The vehicle for the Causal or Higher Mind. The Buddhi is the central core or central point in the Higher Mind. It has its homeplace center in the Indestructible Atom in the middle of the forehead. Buddhi is an ignited and illuminative understanding which produces mentation or interpretation of dreams and visions. Buddhi also pertains to the Informing Principle, igniting the mind with prophecy.

CAUSAL BODY—Higher Etheric Body. The Everlasting Body.

CENTER Q or QUELLE—The subconscious mind seated at the base of the skull is the lower Quelle, meaning psychic womb or source. The higher Quelle relates to the higher unconscious, which is the matrix for the higher mind, the soul and the Higher Self.

CHAKRAS—The chakras are energized, whirling vortices of consciousness. Each chakra is a revolving, vibrating portion of the soul. In all persons, there are twelve chakras strung upon an eternal thread. This thread is called the *Kundalini*. Seven of the twelve chakras may be described as supersensible vehicles for relativity and the phenomenal aspects of the life of the body and the life of the soul. Beyond the seven supersensible chakras, the five superconscious chakras are diamond soul-centers or spheres of eternality.

CHRIST—The Son of God, the Christ, is in command of the Great Archetypal Blueprints in the Kingdom of God, and works to give man the Great Blueprints, Ideas or Archetypes. The Spirit of Truth or Logos works with the Christ; thus, the Christ is known as the *Word*. Since the coming of Jesus, the Christ has centered His Light into the core of our earth and has command of the axis of the earth. The *Christ Consciousness* is opened through the Third-Eye canal. Christ Consciousness is experienced between the eyebrows.

CO-ATOM—To be co-atom to any person, one must be on the same wave length of energy emanating from a sacred atom, as to be co-atom to Jesus one must have an open circuit in his own heart's sacred atom to the heart of Jesus. To be co-atom to a Teacher or a Master, one must have a wave-length open circuit or a degree of light in his mental atoms to the mental atoms of his Teacher or Master. In this way he is telepathically communicable to the mind and thoughts of his Master and Teacher, and thus receives a continued flow of instruction. One can develop his own capacity in co-atom association through meditation and through thoughts of oneness with all life as given of God. When two persons have instant recognition of one another and absolute congeniality, they are co-atom to one another. This can only occur when one has been with the other person in many lives where relationships have proved to be harmonious. When God prepares to use a person for a greater work, He first sends to the one chosen a Teacher or Master; secondly, He sends to him a co-atom person in the world, that the work may be consummated and fulfilled.

CO-DISCIPLES—Co-disciples have shared a spiritual life in former lives and now work in both the physical world and invisible world in pure alignment as catalysts to one another. Such co-disciples are in telepathic accord to one another and have one aim: that they may become like the Disciples of Jesus so that they may serve in the world.

CONSTANT—The Constant is located in the Divine Thalamus above the medulla oblongata in the back of the head slightly below the crown of the head. This is the dwelling place of our Father and the dwelling place of one's Constant. God as Will, God as Life, God as Light, God as Love in the Constant reveals the perfect hierarchy son-of-God to be.

COSMIC—The Eternity System collective consciousness.

COSMOS—The Universal Consciousness.

COSMOS-DISCIPLES—The Cosmos Disciples are the most highly evolved persons on earth and at all times are in direct alignment with the Archetypal World or Kingdom of God. Upon the shoulders of these 33 sacred persons rests the responsibility of telepathic co-ordination and mediation among the 13 Telepathic Disciples, the 3,000 Advanced Disciples, and the 300 Good Brothers living in the physical world. The 33 Cosmos Disciples and the 13 Telepathic Disciples in the physical world are mediators for the Masters. Sometimes, one, two, five or seven are centered within the religions of men, their faiths, and their governments. The telepathic instruction of the Masters is channeled through the inspiring works of the Cosmos Disciples and Telepathic Disciples, and men of the earth are freed from the decaying debris obscuring the Light.

COSMOS-GENESIS—After man has reached the perfect Self-Genesis stage of evolvement, he will become a Cosmos-Genesis man. His emotional body will be fully developed and he will be at one with the love-atoms of Jesus, the Lord of Love. All great Bodhisattvas of the East had reached full development of their emotional bodies, and thus gave to man the bhakti love instruction. John the Beloved, the disciple of Jesus, had a perfected emotional body, and therefore was the closest disciple to the heart of Jesus.

CONTINUUM SACRAMENT—From a continuum of the Bread and the Wine comes the perpetual Epiphany. A Continuum Sacrament, after remaining on the Home Altar or on the Chapel Altar for a 24-hour period, is eaten each morning on arising. To awake to the Continuum Sacrament is to enlarge the circumference or sphere of Eternality. To be aware of the Jesus-atoms in the Bread (body) and the Wine (blood) within the Sacrament is to remain in Jesus as the Way, the Truth and the Life.

DANDA—Universal Law.

DARSHAN—The radiance light exuding from a pure sage, prophet, teacher or avatar.

DEVACHAN—The Fourth Plane of the Astral World. The dwelling place of the Shining Ones, or the Greater Devas. The Devachan Kingdom exists in the First Heaven as well as in the Lower Astral World. In the lower astral devachan planes, the Dual-Devas work with the Cherubim Angels upon the lower chakras to develop man's psychological nature, and thereby produce his birthright of individual creativity.

DEVAS—Co-workers with the Angels in the mathematics of holy and lawful intricacies. The lower devas work with the psychological nature of man. They thrust their symbolic images onto the minds of men through the tunneling action of the subconscious. The lower devas are guru-like but are not gurus. They are an automatic arm of Divine

Mother imposing the Correcting Principle upon man's sense and sensual receptivity. Always with the Archangels are the Shining Spirits or the Greater Devas. The Great Devas are equal to the Archangels. They know God and work as His ambassadors. The Devas are the royalty of the Devachan world. They overshadow large, virginal lands and forests.

DHARMA—Law, Religion, Truth, Virtue. The word *Dharma* means "that which upholds." To be a guardian of the Dharma is to be a lesser saviour bringing salvation or "the Way."

DISCIPLE—The disciple is one who has reached the stage of selfless dedication. He is teachable, having relinquished his egotistical shell. He dedicates, that he may be instructed.

DIVINE MOTHER—The Mother of the World; a Divine Being representing the Feminine Principle. All women come under Her protection and guidance. The Divine Mother works closer to the Lords of Karma than any other Being in the Spiritual Worlds. Mary, the Mother of Jesus, and Kuan Yen of ancient China worked directly with the Mother of the World. The women Saints of Heaven and also the women Adepts, called *Fountains* in the Higher Worlds, work with the Mother of the World. Those who have the grace to behold or see the Mother of the World describe Her as being veiled. It is said She places Her veil over the unknowing and awkward errors of Her earth children. Forgiveness and mercy are the tender theme of the Mother of the World. If one removes himself from the beautiful, firm Laws of God, the Divine Mother initiates him through humility-lessons. Divine Mother comes in with a vengeance when one soils Her chastity.

DIVINE THALAMUS—The etheric counterpart of the physical thalamus or its portion of the nervous system related to the Father in Heaven.

DWELLER—Accumulated karmic entity, reflecting the karmic states of an individual, a family, a nation or a religion. The Dweller must be mastered before one can extend into superconsciousness.

EFFLUVIA—An intelligible, animating, reflecting side of ether through which one is informed. Seership would be impossible without the effluvia chemical action in ether. Effluvia reflects what is, and is an instrument for all pictorial life, living and dead. Clairsentient psychics are familiar with effluvia's chameleon-like and versatile action. Everything a man touches leaves a revealing effluvia through which a psychic may extend his senses to gain knowledge of persons or objects. Effluvia is not a permanent substance, lasting only from sundown to sundown. Ether remains in environments; however, with time the effluvia is withdrawn. One therefore can be aware of ether imprints long after a person has ceased to inhabit an environment; however, if effluvia is absent, he cannot penetrate the most intimate details concerning a person or an environment.

EGO—The higher individuality supported by the higher thoughts of past lives and of the present life.

EGOTISTICAL SHELL—Condensed electrified ether covering the ego, causing one to be bound into the lower mind supporting self-aggression and assertiveness. The egotistical shell has been built by acts of egotism over a period of many lives. This shell must be dissolved through meditation and spiritual works, that the higher ego may come into its own, and the soul be given its fullest expression. One cannot be free from his heavier karma until he has dissolved his egotistical shell.

EIGHTH SPHERE—The great abyss where Satan and his dark angels hold reign. One must cross over the third abyss or eighth sphere when he is being initiated into the higher mind.

EMBLIENCE—Emblience is the womb-like state of lesser ether overdirected by the World-Soul Atom, Race-Lords, Archangels, Angels, and Tribal Species in the period of Nomadic-Genesis. Emblience is recapitulated in the beginning of each Moving Archetype that lesser progressive egos may rise to a higher spiral.

EMOTIONAL BODY—The emotional body and the astral body are one and the same. In the lesser evolved, the emotions work primitively. The emotional body is an ovoid sphere of feeling. The shape of the emotional body determines that every positive or negative feeling shall return to the one who feels it, also that the desires of a person shall inevitably be manifested. The emotional body and the astral core work as hand in glove. The astral fiery core keeps alive and supports the feelings within the emotions and desires. The emotional body is the body through which one enters the door in his seeking God.

ENTITY—An earthbound-dead person lingering in the lower astral planes.

ESSE—Works as a Holy Wind or Gas. The Spirit Essence in all; the Intelligible Divinity-Energy assuring perfection to all, created of and through God. The Esse is the movement of the Holy Spirit in the soul of man, which makes him divine in the midst of his own unknowing. When used as a mantra, produces a state of consciousness called *the Omniscient-Wisdom*. To be free in Esse is to experience purity as Light, and freedom in soul movement and body. Esse is joy liberated into consciousness. Esse produces God-awareness in the physical and in the transcendental states of consciousness.

ETERNAL SUSTAINING ATOM—The supreme atom of the eternal atoms of man. The Eternal Sustaining Atom is the means through which one moves from one eternity to another. All that the soul has experienced in former eternities, all that has been earned, is encapsuled and retained in the Eternal Sustaining Atom.

ETERNITY SYSTEM—Any system having a Sun, Earth and Planets. There are countless Eternity Systems in the Universe. All are born and die as man is born and dies.

ETHER—Not to be confused with physical ether as termed by physics. Ether consists of a life-substance yet unknown to man in his physical state. It is a means by which man transforms all energies. In the physical body and action, ether consists of fire extracted from the Sun. In the spiritual, it consists of the supporting essence of the Universe known as prana and akasia. Ether, the life substance supporting all life, is called *prana* in the East, and called by some *bioplasma* in the West. Ether is a coagulating semi-gelatile and semi-fluid substance. Ether is quasi-tangible. It can be photographed. Ether supports electric and magnetic action and other forms of energy yet to be discovered by man.

ETHERIC BODY—The double of the physical body. It is made of ether and prana. Its life substance is supported by the Sun. The lower aspect of the etheric body supports the life in the physical body and the life of the lower mind. The higher aspect of the etheric body supports the spiritual and higher mental life. The higher etheric body survives death. The lower etheric body dissolves with the physical body.

FAMILY-ATOM—An etheric encasement psychically charged. Father, mother and children in a Family-Atom are held together by the psychical charge or lines of force which have attracted them to one another. The low charge of psychic energy in a Family-Atom keeps alive the soul-debts memory between persons born in the family. The combined soul-grace of the persons born in the Family-Atom is watched over by a Family-Atom Guardian Angel. If grace is abundant, all souls in a Family-Atom progress. If the Family-Atom is heavily laden with karma, the result is suffering and sacrifice, that all may eventually evolve in a cluster of human souls.

FAMILY-ATOM DWELLER—The condensed negative shadow-body of family karma challenging the souls encased in a Family-Atom. The sins of the fathers are visited upon the children through the family-dweller action.

FAMILY-GENESIS—Persons dependent upon ancestral myth inheritance as expressed through a mother and father in a Family-Atom. Family-Genesis impulses seek to build a society patterned after ancestral heritage. From the Family-Atom compulsion comes the building of churches, the building of societies and education.

FATHER—Our Father dwells in the Third Heaven. He centers His action onto the Earth through the World-Soul Atom, known in the Bible as "the bosom of the Father." (St. John 1:18) The Father works with the Will in God that man may (1) perfect a physical vehicle or form, (2) expand the image-power of his soul, and (3) command the powers of substance in this eternity. Direct Hierarchy-knowledge comes to man through our Father, who shapes and forms all life in this eternity system through His *Constant* which has imaged man. God as our Conditioned Father is a Being. God Absolute is Spirit. Our Father in this eternity system, to whom men turn as His children, seeks to bring them to Absolute God and make them sons of God.

GANESHA—In Sanskrit, means Supreme Essence. Ganesha, a great Deva, the son of Shiva, was also Ptah of Egypt and Hermes of Greece. He is the Deva over industry and literature.

GARUDA—Powerful myth-bird Deva-god or man-and-eagle combination representing freedom in earth and in space.

GOD—The Eternal, the One, the Creator. God is Being within Being. All galaxies, nebulae, worlds, eternities, systems, are embodied in Him. From the smallest cell to the greatest star, all Creation is God. God as Spirit is Intelligible Energy. God is the Absolute God; the Father is the Conditioned God.

GOOD BROTHER—There are three hundred Good Brothers working at all times in the earth. Their mission is to appear as a means of help and succor when all else seems impossible. In their physical memories, the Good Brothers rarely know they are Good Brothers; in their etheric memories they know, and are compelled to act during crisis periods. In time of trouble, they smooth the way and move on, asking nothing by way of reward.

GRAND-PSYCHE—The Conscience.

GREAT IMMORTALS—(see Masters)

GUNAS—The three energy or force vehicles through which man develops; denotes a quality as well as a force. These vehicles are called *Tamas* (inertia), *Rajas* (electrified action) and *Sattva* or *Sattvic* (peace).

GURU—Teacher.

HALL OF LEARNING—Located in the seventh plane of the First Heaven. All initiation in the Hall of Learning is preparation to serve as a night healer and a daytime initiate. Dream research of the night is under the supervision of the great Masters and pure gurus in the Hall of Learning. All spiritual aspirants are initiated in this precinct of Heaven, that they may render a knowing service in the world.

HALL OF WISDOM—(see Heaven, Second)

HANUMAN—The monkey Deva-god psychological aspect of the mind.

HEAVEN, FIRST—The First Heaven is the Higher Astral World. Each of the planes of the Astral World contains seven regions. The lowest, or Plane 1, is the Grotesque plane. Plane 2 is Fantasy. Plane 3 is Wish. On the Fourth Plane of the Astral World, the action of the First Heaven begins. Here one reads his akasic records and hears the Audible Sound or Music of the Cosmos. On the Fifth Plane of the Astral World, one unites with the Masters, the Saints, and begins his night instruction and night-ministry work. On the Sixth Plane of the Astral World, one enters into the prophetic side of dreams. On the Seventh Plane of the Astral World, one contacts the great Bodhisattvas who prepare him for initiation into the Second Heaven.

HEAVEN, SECOND—The greater Cloisters of Heaven where one meets the Most High Saints, Mary and the Great Masters, and receives instruction in the Hall of Wisdom or Shamballa. Adepts are initiated in dreams through the seven planetary Logoi or the Archangels over seven planets. These prepare the adept to make union with the Third Heaven.

HEAVEN, THIRD—The Third Heaven is the homeplace of Jesus and His Apostles. Here, the initiate and adept receive the power of Archetypal Light directly through the assistance of the Christ Spirit overdwelling Jesus.

HIERARCHS—The Elohim, the Host or the Zodiacal Overlords assisting this earth or eternity system in its development. The Hierarchs are the Mediators between the galaxies, the starry bodies and the planets. Man's creative genius is inspired in regular intervals through the uplifting tide of creation translated to him by the Hierarchs. Hierarchy uses the power of imaging or making. Sending their Rays into the Sun and the Earth, they assist the Father and the Christ in the creation of mankind. Each Hierarch Overlord is a zodiacal prototype or blueprint for man, such as Aries, Taurus, etc. Hierarchy, using Time and Space through the power of the Constellation Rays, works with the Will in God that each man may perfect his own prototypal likeness of Hierarchy. *"Let us make man in our image, after our likeness." (Genesis 1:26)*

HIERARCHY NATURE—The Hierarchy Nature is experienced as consciousness in Buddhi, which is the Center of the Informing Principle in the center of the forehead.

HIGHER SELF—The Higher Self has its dwelling place in the soul energy-field above the head as far as the hands can reach and touch. The Higher Self is centered within the Eternal Sustaining Atom, which is the vehicle for the crossing over from one eternity system to another. The Higher Self contains the image of what man has perfected in other eternities or world systems. The Light of the Higher Self may be experienced through prayer, contemplation and meditation.

HIGHER UNCONSCIOUS—The higher unconscious is primordial purity, containing flawless intelligible truth. There is no contradictory or controversial functioning or phenomenal functioning of this primordial purity. It is the *Esse*, or the *Essence* of Eternal Mind functioning as Buddhi in the higher mind.

HOLY GHOST—The Holy Ghost works with the Will of God, the tones of the earth, the movement of the earth, the vibratory velocities of the earth, the pulsation of man's soul, the speech of man, the music of Heaven, and all of the sounds sounding and resounding in the earth. The Holy Ghost commands the tone and sound in all living forms. The Holy Ghost is a Being. Holy Spirit works through the Holy Ghost, as Christ works through Jesus. The Comforter, the Spirit of Truth and the Holy Ghost are one. (St. John 14:16,17,26)

HUMAN-GENESIS—Same as Family-Genesis.

ILLUMINATI—One of the higher orders of the Risen Dead in the First Heaven. The Illuminati and the White Brothers work in unison. The Illuminati maintain an inner school of immortal ethics. All Illuminati reincarnate. As geniuses in the earth, the Illuminati are selflessly devoted to create, that mankind might rise and become God-men through creation. The Illuminati incarnate in timing to the need of the souls of men, that they may inspire men to create and to serve God.

INDESTRUCTIBLE ATOM—A Bija point in the center of the forehead called *Linga-sarira.* This is directly connected with the akasic-record recovery. The Indestructible Atom in the middle of the forehead was opened for man by Jesus when He ascended. Due to the quickening and emergence of the Indestructible Atom, the mentality of man has undergone a tremendous transition within the last 2,000 years. The Indestructible Atom enables man to receive the Power of the Holy Ghost; it will eventually enable him to image with manifestation powers.

INITIATE—An ego who has followed a spiritual path in previous lives and is in a state of being initiated into greater illumination and spiritual power in this life, that he might better serve the world.

JAPA—Repetition of a sacred mantra or name through the use of a mala or of certain mudra postures.

JEHOVAH—The Race Lord working with the Moon. He is the Over-Lord of the second chakra. All persons encased in tribal development are under the direct influence of Jehovah, that the race may become a purified vessel for the Human Race. In India, he is known as Indra, the God of War. Jehovah has control of the reincarnation cultures. In one lifetime, one experiences many phases of genesis culture. Jehovah works through war to produce race, to preserve races, but he also uses, with a ruthless hand, the destroying of races. His work is a necessity-positive through the use of the negatives and positives in race. Jehovah is in command of 33 devachan heavenly precincts. He will influence the left side of the brain in man until man masters the instinctual laws experienced in Tribal-Genesis. One has to move beyond the 33 precincts of Jehovah in the lower astral regions, that he may reach the Second Heaven where dwell the true Archetypal formats for all religions in the Earth.

JESUS—The Sovereign Mediator between the world of man and the World of God. The Regent of this eternity system. The World Saviour and Messiah. The Son of Man, Jesus, dwells in the Realm of Light with His twelve Disciples. He is at one with the Father and is the most direct Mediator between man and God. Through His many Saviour trials in various bodies, He made it possible for the Christ Spirit to penetrate the core of our earth and to command the axis of the earth that the Christ Light and Christ Mind might come into the world or that man might receive the Greater Archetypes as conceived in the Will of God. Jesus is in command of the higher etheric body of this eternity. He makes it possible for men to become mediators unto one another, ". . . *love one an-*

other." (St. John 13:34) He also makes it possible for man to enter into mediative relationships with His Cosmos Family of the Spiritual Worlds. Jesus works now with the heavy karmas or burdens of men and with all angelic kingdoms. He works with Jehovah that men overcome the crystallized tribal retributive laws; He also works with Jehovah and the Propagation Angels to overcome the racial compulsions set up by the World-Soul Atom so that men in the world may eventually become free of race.

While in earth, Jesus' alignment with the Christ enabled Him to use the Archetypal Powers, and He thus saw in all dimensions; He was aware of the destiny of the world and of man. Through His relating and alignment with God, He had knowledge of the Eternals. "*. . . glorify thou me with thine own self with the glory which I had with thee before the world was." (St. John 17:5)* Through His alignment with greater Hierarchy, He understood that men are yet to reach prototypal perfection. Therefore, He could say on the Cross, *"Father, forgive them; for they know not what they do." (St. Luke 23:34)* Jesus worked through the Triad action of the Father, Melchizedek, and the Ancient of Days, Sanat Kumara; thus He had the power to change one substance into another (water to wine), to multiply substance (fishes and loaves), to change one energy into another (healing miracles), to disappear and appear, to command the elements, and to raise the dead. The Lord Jesus is yet in alignment with God, the Christ, Hierarchy, and the Triad action of the Father. He now works that man may make this alignment and eventually do these and "greater works" (St. John 14:12), for the etheric condition of the earth will change, thereby enabling man, in his future atom evolvement, to do "greater works." Man will also reach the time when death will be overcome and *"there shall be no more death." (Revelation 21:4)*

Christ Jesus is the promise of the Constant in God. Christ Jesus is the Mediator between and for all Teachers, all Gurus, all Sages, all Bodhisattvas, all Saints, all Mediators, all Adepts, all Masters. *"This is my body,"* said Jesus. Men are embodied in their soul-natures within the embodiment of Christ Jesus, where there is no separateness or division or the need for the casting out of the perishables.

JNANA—Mind-wisdom consciousness.

KALI—The dark aspect of the feminine Shakti instinctual power, spoken of by Carl Jung as "the terrible mother." A kali woman may also be unmarried; her sense of amorality makes her a threat to both sexes, male and female.

KALI-YUGA—The ending of an age in which the heaviest karma is present.

KARMA—To the Western world is inclusive of the Law of Sowing and Reaping, the Law of Retribution, the Law of Consequence affecting all actions, good or bad; the Law of Cause and Effect.

KARMA-YOGA—Work service as an offering to God.

KLOSHA—Veil, mist, psychic web, demoting mental spirituality. The Klosha veils prevent man from having direct union with God. Through mediation and work-service for God, man must dissolve the Klosha veils.

KRISHNA-POINT—The Krishna-point is the Mouth-of-God Center; it also relates to the 11th house in one's zodiacal chart. In the 11th house may be seen how one relates to the Masters and their instruction.

KUNDALINI—The Kundalini has been called the Serpent fire because its action is serpent-like. It is also spoken of as being electric because it is the major electromagnetic conduit for the energy intelligible processes in matter, in life. It is also the vital connecting vehicle between body, soul and spirit. When unawakened, Kundalini is the sleeping, intelligible Royal Vein circulating the soul processes into all phases of life and existence. Through awareness of its action, the kingly science of soul union and conscious awareness with the One, or God, is made possible through this Royal Vein or fiery Serpent.

LIGHT-STREAMS—The seven Light-Streams are the stepped-down Rays from Hierarchy. The Hierarchs centered in the Constellation-points send their Rays into the Solar System. On penetrating the zodiacal system, the Rays of Hierarchy, through the Mediation of Jesus, become the Light-Streams which shape and form man. Presently, man can incorporate only seven Light-Streams. He will incorporate the other five Light-Streams when he becomes like Jesus.

LOGOS—The audible sound of the Holy Ghost speaking through inspiration, illumination and revelation; divine speech or descriptive verbalism experienced and inspired in spiritual illumination. Sometimes referred to as a center in the throat. Between the eyebrows may be found its corresponding center. These two centers are the illuminative points through which one expresses the Holy Ghost, and thus contacts the Greater Archetypes. All Great Teachers have these great centers or doors open continually.

LUNAR BRAIN—The abdominal automatic primitive brain supporting the instinctual life. The center of clairsentience situated in the solar plexus.

LUNAR PSYCHIC—One who is engrossed psychically with his emotions and sees all through emotions and feelings. He is dependent upon the astral lunar reflective light for his psychic powers. The lower lunar psychic is unable to interpret what he sees. The higher lunar psychic sees in part.

MAHA CHOHAN—The Being in command of the Third Light-Stream. (Order) The Lord of the Continents and the Lord of World Governings. His work for the earth is activated every 3,000 years and is sustained for 300 years. His etheric proximity to the earth creates major tumults and upheavals among the humanities of the world. Decadent rules and corrupt policies in governments come to an end; fixed traditions in societies and families change; and limited philosophies in religions undergo metamorphosis and transition.

MAHA-MATRA—Master-woman under the Divine Mother. Instantaneous healing or God-Realization.

MANAS—Mind.

MANDALA—In Sanskrit, Mandala means *to be in possession of one's essence. Manda* (essence) + *la* (to be in possession of). Sacred circle.

MANIPURA—In Sanskrit means the *fulness of jewels.* The Third Chakra or Solar-Plexus Center. Manipura as the Third Chakra is the center of the emotions and desires; the element of Manipura is fire.

MANTRA—*Man* (thinking) + *tra* (protection). *Mantra yoga*: the sound-vibration in the use of the mantra provides protection.

MANTRAM—In Sanskrit *Manas* means "mind"; *tri* means "to cross over." That which enables one to cross the tempestuous sea of the mind. The speaking prayer in which the vowels and combining of words build the power to align one with the spiritual helps within the Spheres and Realms of Light. The sounding of word-combinings to dissolve karma, tension and fear. A mantram is a formula of words containing Bija or seed-sounds of superconsciousness. In every mantram is hidden a mantra as a master idea and sound providing physical mastery and spiritual experience. A mantram contains molecular energy-particles of light. A mantram spoken with love and absolute belief is a freeing way.

MANU—Law. The great *Manu*, or Legislators of the Law, gave the great allegoric-creative techniques of the Universe. They came before the Rishis. The Manu or Lords of the Law work with the Seraphim or the Lords of Mind called *Manas*. In Egypt, *Ptah* was the being who represented the Manu. Ptah and Ganesha of the Hindus are one and the same. All great Scriptures are over-directed by the Lords of Manu and Manas.

MARKING AND TRACING—The first stages of initiation are *Tracings*. These are one's chief negatives he must work with when he starts on the Path. Markings are divine signs and grace-reassurances experienced during illumination. When one receives a marking, he has attained a station in Light through which he will serve as a pure and whole channel, that God may use him. One moves from marking to marking as he evolves in God-Realization. An open or uninhibited character expression is manifested by Marking and Tracing. Marking and Tracing techniques enable one to experience his frame of consciousness within the Greater Archetypes.

MARY—The most chaste of women in the earth, chosen of the Father to give birth to our Lord, is present with men as Her Son, Jesus, is present. The Mother of Jesus works with the mothers of the world. She does not re-embody or return to the physical world. Mary works to give one Chastity so that he can receive the Immaculate Conception in the heart. Mary works with the Most High Saints and the Mother of the World. Joseph and Mary work together as co-atom Cosmos Souls in the Second Heaven that their Son, Jesus, may enter the heart of the one who prays and makes supplication to God. Joseph and Mary are conjoined in one work together in command of the religious impulses—regardless of what a religion is called—as long as Jesus is presented therein as Christ Jesus.

MASTERS—Advanced Beings of high degree dwelling in the Second Heaven or Spheres of Light. The Masters, also known as the Great Immortals, have retained cognizant and clarified memory of other eternities. Having lived in other eternities similar to our earth system, they have mastered the opposition forces yet to be mastered by man in this earth. The Masters are not to be compared to personal gurus or teachers called "masters" in certain parts of the world. Until the year 1946, the Masters or Great Immortals inhabited physical bodies only on rare occasions. These immortal men will no longer re-embody in the physical world, for they have been replaced in the physical world by 33 persons called Cosmos Disciples. The Masters have been men; they are living vessels to inspire immortal and deathless thoughts. They are greater than Initiates and greater than Adepts. While Jesus is the Mediator of the heart, the Great Immortals are the mediators of thought through telepathy. The Archetypal Masters are presently overdwelling the light, life and the souls of those who are called to teach, to heal and to reveal telepathically what the Greater Archetypes are saying to men of the Earth. The Masters are the divine collaborators with men. Their works are to intensify the skills within the soul-cultures of men, that men might identify or align themselves with the Light of the Christ, the Love of the Father and the Will of God.

MASTER G.Q.—The Little Brother. The one called G.Q. is spiritually close to the Eastern, yogi initiate, and also to initiates in the Western world. G.Q. equalized the flow of ancient Archetypal ideas. He is the last of the true Masters even though considered as the least of these. The Western initiate, being unable to assimilate the yoga practices and formulas, is enabled, through the Master G.Q., to retain the basic truths within the ancient Archetypes of the Vedas, the Sutras, the Upanishads—for example, the law of birth and rebirth or re-embodiment. The laws of eternity justice, or karma, are made clear to the Western initiate through the work of the Master G.Q. He works with the disciple of the new age to retain the old and, thereby, to gain the whole in Christ.

MASTER HILARION—The Master Hilarion or Saint Paul is in command of the Fifth Light Stream (Science). To unite with the value of the full measure of the Fifth Light-Stream is to come under Saint Paul. Saint Paul of Tarsus was a scribe-apostle. He knew the Christ as no one else knew him. His Epistles in the New Testament of this era's Moving Archetypal Bible were composed directly through his Archetonal Grace or of hearing the Word and of speaking it so that it lives now in a soul-vernacular so close and intimate that in reading or speaking his words, Soul-Realization is made possible; communication becomes healing of spiritual ignorance.

MASTER K.H.—The Great Immortal called "K.H." in the Higher Worlds was Hippocrates, the father of modern medicine. This Great Being will no longer re-embody in the physical world. He is now in command of the Fourth Light-Stream. (Law) Through Him, men overcome oppositions, agitations, quarrelsomeness, irritation. It is also through Him that men become peacemakers, servers, and healers of the nervous, muscular, circulatory, cellular, and blood systems of the body. Master K.H. works with Sanat Kumara to enable man to attain a perfected and symmetrical physical form. As the Master K.H., the body used last in earth, he assisted the changing of the mystical currents in India that the Eastern teachings might be moved through the Light of Buddha onto the Christ into the Western world.

MASTER M—Also known as Master Morya. Master M is in command of the First Light-Stream (Will). Those who come under His telepathic instruction are called will-initiates. Master M was Plato in a former life.

MASTER R—Master R channels telepathy through the Seventh Light-Stream (Ritual). He holds the key to the survival of religions in the world. He also enables the dedicated disciple to make synthesis, or to blend with all seven telepathic Light-Streams of the Masters. In one of his previous lives, Master R was Hiram Abiff, who assisted Solomon in his initiation. In the Higher Worlds, Master R is called the Great Phoenician. In China, he was Lao-Tse. In Jerusalem, he was Joseph, the father of Jesus. The mediative telepathic power of Master R in the present time relates to the dissolving of crystallized ritual and forms in worship, and overcoming of bigotry in religions. Under the telepathic tutelage of Master R, initiates are enabled to undergo the *molten sea* initiation within the core of the earth, and to arise above the so-called 8th sphere or the ring of *outer darkness* around the earth. Through the work of Master R in this age, new musics will come into the earth; the ability to read the records of past lives will become more prevalent; and a new ritual for worship will be established in religion which will come closer to the true ethic in worship as established by the Lord Jesus.

MASTER SERAPIS—Hierophant of the Inner Pyramid. The one called Solomon in the Bible. Master Serapis over-dwells the Third Light-Stream (Order).

MAYA—The changing world of gravity energy producing change, creating the illusion in man that earth life is all. The Sanskrit meaning is *to measure*; illusion. Maya is karma; it is a mathematics system in density.

MEDIATION—Mediation is the most unselfish and unclaiming means of spiritual serving through which one remains impersonally involved with the karma of those whom he would heal or help.

MEDIATOR—One who makes himself a divine artery or channel for the Light. A mediator asks for no rewards for his mediative prayers and suggestible helps. He only asks to remain a perfect instrument, that the Power and Will of God may flow through him, supporting, healing and lifting. When the mediator is sincere and wholly dedicated, he is free from the karma of those whom he would heal and help. He avoids boastfulness of his healing works. He asks no personal or physical rewards for his healing helps, knowing himself to be but a channel through whom God sends and heals. The highest technique used by a mediator is his use of Angel-to-Angel Mantrams, as he knows on his releasing the one whom he would help to the Angels' suggestible helps, that the Angel taking charge of the one to be healed knows with an exact and precise wisdom what can be done and what will be done. Thus, in the angelic mediative helps, miracles occur, as it is left to God to reveal to the Angels what is the right and just way for the healing to come.

MEDITATION—Rightful meditation draws one into higher and finer rates of vibration where one's thought and emotions are changed to superconsciousness. As a true meditator, one becomes a steward of time and of will. When one accepts his placement and destiny in the era and time—acknowledging himself to be a householder

—he becomes both spiritual and practical, reaping the fruits of a spiritualized life. And with each day's progress, he comes eventually to union with God, which is the ultimate in meditation.

MELCHIZEDEK (MELCHISEDEC)—Mentioned in both the Old and New Testaments. Melchizedek works with the Great Immortals within the Spheres of Light to teach the initiate how to overcome the four elements of the Earth—fire, earth, air and water—that the initiate may obtain the power of Manifestation and De-Manifestation.

MENTAL BODY—A composite field of light. The higher mind expresses itself in the mental body through three mental atoms. The lower aspect of the mental body, called the lesser mind, is dependent upon the psychic energy coils of force inherited from past lives, from ancestral mental habits. The lower mind serves the physical senses, using instincts from Tribal-Genesis memory and ancestral memory. It is the work of all spiritual aspirants to still the more atavistic aspects of the lower mind or lesser mind and make of it a complementing additive partner to the Higher Self and the higher mind. The higher mind seeks to clear the field, that it may come forth as a supernal instrument in creation. In all selfless works of creation, the higher mind is in command.

MOUTH-OF-GOD CENTER—The medulla oblongata at the base of the skull, in proximity to the brain stem, is the *Mouth-of-God Center* or the Logos Vibration of the Holy Sound.

MUDRA—A posture of hand or fingers used to open the seventy-two thousand nadis in the etheric body and reach the higher stages of meditation. Behind the system of the gesture of the hands are the Great Laws.

MULADHARA CHAKRA—The First Chakra seated at the base of the spine; its element is earth. Muladhara means "root support."

NADAM—Sounding the Name of God.

NADIS POINTS or CANALS—Seventy-two thousand energy stations in the lesser etheric body.

NAGAS—Fire Initiates with Siddhi Powers are called *Nagas* or Serpents. The Bible speaks of them in the passage: "Be ye therefore wise as serpents, and harmless as doves." (St. Matthew 10:16)

NESCIENCE—Ignorance, unknowing.

NEUTRAL FIELD—A non-decisive mental atmosphere inviting entity possession. A neutral field is the product of many lives. One builds a neutral field by refusing to participate in the responsible issues of life. A neutral-field mind is described in the Bible as being "lukewarm." It is said that persons having such minds are "spewed out" and of little use to God. Mediumistic powers in the hands of a neutral-field person are sometimes pure, sometimes impure, as both the dark and the light can be housed at one time in a neutral-field mind.

NIBLIENCE—Niblience is the working of the World-Soul Atom, the Hierarchs, Race-Lords, Archangels and Angels who in their Archetypal combinings work as an overshadowing, sending forth and determining the Species within the period of Human-Genesis. These combinings move through the lesser planetary rays to build that of family on a spiral higher than tribe.

NISCIENCE—Knowing. Knowing beyond knowledge. Superconscious Knowing; the memory of the Eternals. Niscience is the hidden side of science or that as yet withheld from science. Niscience is knowing beyond knowledge or a knowledge gathered out of the Eternals to be recovered in Self-Genesis.

NOMADIC-GENESIS—Same as Tribal-Genesis.

OJAS—The sacred life-force centered in the sexual drives to be raised to the heart center through sacred thinking during the sexual act, thereby healing and eliminating lust. Creative sexual energy as a vehicle for the Holy Spirit. During the sexual act, to lift the sexual force beyond the lower chakras through dedication to God is the aim of all who would attain the polarized or androgynous mind. Ojas is a form of ecstatic willing during the moment or second of ejaculation or orgasm. The sexual energy as Ojas flows upward to the heart and the inner etheric brain of man, whereby he is given a deathless vitality of will and mind.

OM—The OM sound is both prayer and mantra. OM erases the negative past, explodes the present negatives, creates the future as perfect, as joy.

OMNISCIENT CELL—In the etheric brain is developing an Omniscient Cell. The Omniscient Cell in the brain is the Jesus Cell. Men will be ultimately like Jesus.

PRAKRITI—Supreme Nature. The Divine Mother functions as Prakriti.

PRANA—Life-force energy. The sum total of all the manifested and the unmanifested energy in the Universe. All of the forces of nature, heat, light, gravitation, electricity, magnetism, the atom and atomic energy are aspects of Prana. This energy springs from Absolute God. Prana is the higher energy level of ether working simultaneously with the molding and shaping action of the effluvia in the ether. The energy in high prana has yet to be analyzed by science. One contacts prana through breathing. Prana life-force may be unlocked and freed into the etheric body through cosmic exercises, yoga, breathing, speaking of mantrams.

PRO-GENESIS—When men have become like Jesus, as promised in I John 3:1-3, they will be Pro-Genesis men with cosmic powers of manifestation. They will do all things, as did Jesus. Following Pro-Genesis will be *All-Genesis* in which all mankind will be at one with our Father which art in Heaven. And finally, in the period called One-Genesis, men will become the sons of light with hierarchy powers.

PURUSHA—Soul-wisdom. Union. The Father Principle as Soul Consciousness.

QUELLE—(See Center Q)

RAJAS, RAJASIC—Extreme or fiery energy; over-accelerated tensions in emotions and mind. Heated desire; greed; selfishness; egocentric action.

RISHIS—Ancient sages, wise men and teachers of the East. The Rishis are Elect beings with powers correlating to Hierarchy. They originally came as Source souls from the eternity-system cluster of the Great Bear. The Rishis bring the *Laws* for interpretation of karma. Their present station is the Second Heaven with the Logoi or Planetary Archangels who overdirect the reflected spheres of the planets.

There are seven great Rishis who gave the *Vedas* and the *Upanishads*; the Abraham thread also includes the Islam reflection of the Rishi. All teachings originate with Rishi teaching. The Egyptians were heirs to the Rishis. The Christ, as the only begotten Son of God, is beyond the Rishi. In the Christ, one coordinates all teachings and laws with love.

RUDRA—The John-the-Baptist power sealed at the base of the skull.

SACRED ATOM OF THE HEART—A deathless atom. Contains the restoring-life sealed into all men by the Father since the beginning of this eternity. Works in a triad action with the Eternal Sustaining Atom and the Indestructible Atom so that man will remain an eternal being while experiencing the prototypal works in the earth.

SADHANA—Initiatory disciplines.

SAHASRARA CHAKRA—The Seventh or Crown-of-the-Head Chakra.

SAINTS—The Saints are holy persons who have suffered many martyrdoms in the physical world. The true Saints remain within the First Heaven and rarely re-embody. The Saints are Mediative Beings. One makes petition to the Saints for personal helps. The entreating soul implores the Saints because they are the first Mediators to the human condition. Having suffered every human circumstance of trial and error, the Saints place all petitions of human suffering before the Lord Jesus. He, as the Mediator between the Father and man, pours down His bountiful resources into the woes and griefs of man. The most heavy sinner condemned by his own sense of guilt, when contrite—calling upon the Saints for redemption—has begun his own pilgrimage toward becoming a Saint. Released from the Wheel of Karma, the Saints now dwell on the sixth plane of the Astral World. Their labor is to extend the prayer faculty of man, that the tone in prayer may be sustained longer; and, therefore, prayers may be answered as to timing devoid of karma.

SAINTS, MOST-HIGH—In the Second Heaven, Mary and Joseph overdwell the Most-High Saints. These Saints are the inhabitants of the seventh or apex Cloister, the Cloister of the Holy Precipitation. The Most-High Saints are celestial anointed Saints. To come under a Most-High Celestial Saint is to free the power of stilling the tumults in the human sensory trials.

SAMADHI — A state of meditation beyond relativity. In Sanskrit, *sam* means "union;" *adhi* means "Lord."

SAMSARA — The Sanskrit word for the phenomenal world; that which is moving intensely.

SAMSKARA-SEEDS — Karmic tendencies brought over from former lives.

SAMYA — One who desires total union with Absolute God.

SANAT KUMARA — Sanat Kumara is one of the lesser Fathers of this eternity system. He works with "our Father in Heaven" to time to man the inner meaning of days. He is spoken of in the Bible as "the Ancient of Days" (Daniel 7:9) When a disciple comes under the telepathy of Sanat Kumara, he is in timing to his destined purpose. The great Sanat Kumara or Ancient of Days initiates the disciple into timing and timelessness in conjunction with one's Guardian Angel and Recording Angel. He is the Initiator for Time-freeing, whereby one is freed from the Maya fixity or limited units of measure. The Ancient of Days, as the Preserver of Time and the units of measure, is in command of all etheric encasements in this earth system. He is in control of timing as to the length of the functioning of a Race, the length of the years in one's life, the length of an Archetype's sounding and sending, and the length of mankind's mortal days. Sanat Kumara works to aid man to command the powers of Saturn, enabling man through heavy discipline to overcome the turbulent forces of the Astral World. He is the Father of the

Devachan light of life in the Astral World. This kingdom works more closely with gravity laws in the Maya system affecting the physical world.

SANKALPA—To make a resolution.

SATTVA, SATTVIC—Law and virtue, giving the bliss joy or peace. Purity; goodness; harmony; rhythm.

SELF-GENESIS—Self-responsible as to conscience. There are two stages: Lesser Self-Genesis and Higher Self-Genesis. In Lesser Self-Genesis, the individualistic person is engrossed with his own self-interest. In Higher Self-Genesis, one recognizes the right of every man to become an identity and to relate himself to the Cause of his being, or God.

SHAKTI—The feminine or subjective polarity.

SHIVA—The masculine or positive polarity. Shiva is spoken of as the destroying principle; his work is to destroy evil.

SIDDHIS—Power attributes of extrasensory perception. The lower Siddhis are psychic. The higher Siddhis are spiritual. All great Masters warn against the use of the lower Siddhi action.

SILVER CORD—The etheric cord enabling one to travel out of the body while awake or asleep. It is anchored etherically over the spleen, the liver, the heart, the throat and the crown of the head. At death the silver cord is ruptured.

SOLAR INITIATE—He is united to the Informing Principle. In his psi powers, he is a scientist working with the Spirit of Truth.

SOLAR PSYCHIC—Same as solar initiate.

SOUL—The soul is centered in its pulsation point within the crown of the head. The soul is a supernatural energy overdirected by the Presence and Image of God. The soul is a vehicle for spiritual experience. It has an *Eternal* mission. To trust in the soul's Eternal mission is to produce ideas of the Infinite. Jesus laid down the Ethic for the life of the soul in the world of man in His Sermon on the Mount and in His reminders of the necessity to live within the Commandments. The *Most-High Soul* is a superconscious energy-vehicle of movement. The Most-High Soul is at all times giving birth to pure creation.

SOUL-MEDALLION—A pulsating vortex of supernal light in constant movement around the head of man, keeping alive his soul impulses and mental creative compulsions. The soul's medallion works in conjunction with the heartbeat expansion and contraction. In the uttermost upper point of the soul's medallion directly above the skull is a pulsating vibratory action. This pulsating action becomes the heartbeat for the spiritual body after death. The soul's medallion records on its outer rim man's negative actions. This is called the *vibratory hum*. The vibratory hum of the soul's medallion is reflected into the lesser mind. Each time a person meditates, he must clear the field of the vibratory hum and slow it down.

SOUL-SHEATH—A rim of incandescent light around the physical body. This light flows from the light of the soul's medallion and is built from grace and purity.

SOUND CURRENT—The four variations of the Audible Sound Current are: *Para*, works with the Will aspect of the Sound Current; *Pasyanti*, works with the Light aspect of the Sound Current; *Madhyama*, works with the Love aspect of the Sound Current; *Vaikhari*, works with the

Life aspect of the Sound Current. The Vaikhari Current is called by some *the Audible Life-Stream.*

STELE—The starry point between the eyebrows, or the center of command through which the higher consciousness moves forth into the world.

SUSHUMNA—See *Archetypal Cord.*

SVADHISHTHANA—The Second Chakra located at the pelvic center.

TAMAS, TAMASIC—Inertia, slow, ignorance, resisting, dark.

TANMATRA—Manifestation powers, producing miracles beyond sense. The energy essence within each element, whereby one unites with the meaning of the elements: earth, water, fire, air, ether, light, tone.

TELEPATHIC DISCIPLE—Every Telepathic Disciple works directly with one or more Cosmos Disciples. The union of minds between these souls is necessary to produce manifestation of the Dharma. All Telepathic Disciples are centered within the Telepathic Matrix on the functional, physical and spiritual levels. They have access to the mind-banks gathered from soul-cultures built by the interflow between the Greater Minds of the Elect who mentally and spiritually direct and influence the human spirit. The Greater Telepathies are open to the Telepathic Disciple especially when he is close to becoming a Cosmos Disciple and therefore a direct revealer of Archetypal Light. The Greater Telepathies are those that keep alight the Light-chain of ideas, passing them onto enlightened souls with hierarchy-nature minds. A Telepathic Disciple has mastered the sensitivities of the senses on the gravity level and is ready to be a vehicle for the Cosmos-Universal Power of Passing. One enters Telepathic Disciple apprenticeship as he approaches his mid-forties. One enters Cosmos Discipleship when he has entered into his latter wisdom-years.

THIRD EYE—The Third Eye dwells directly beneath the soul's pulsation. The Third Eye flows its light of vision outward between the eyebrow center, which is called the Jade Gate or the Cave.

TRANSENERGIZATION—Every dark mass of negativity has one penetrable point of energy. When Light penetrates this point, the complete mass is transenergized. The great Masters, seers, healers, and the Angels use this technique of light-energized power with freedom and with knowing. Any situation of darkness contains its penetrable point. Healing is made possible when one has acquired the power of transenergizing.

TRIBAL-GENESIS—Nomadic segments or clusters of people who have interlocking blood ties sealed into tribal encasements. Dependent upon primitive etheric laws, Tribal-Genesis persons live close to the tribal consciousness and taboos of their forefathers.

UNDERSOUL—The shadowed self. Consists of the lower subconscious and the egotistical shell. The home of all ego reflections which contain all of the mechanical reactionary systems of former lives, ancestry and various personalities used by the ego. The Undersoul collects the debris of wrongdoing, non-doing, and fantasy astrolocity. Undersoul debris produces automated suffering. An overburdened Undersoul creates combustion, then fire, then pain. Faith in God sets fire to the accumulated debris in the Undersoul. Such fire produces purification. The Undersoul presently seeks to build another brain-vehicle which will become the center for the higher mind once it is perfected.

UPA-SOUL—The luminosity-sphere of the soul's vibration around the head as related to the higher unconscious. The Upa-Soul is the Pathway to superconscious energy and mediating transcendence within the Most-High Soul.

VAJRA—An act of the will used by the guru to erase karma in a chela.

VENERABLE ONE—The Venerable One or Brhaspati is a Rishi, a Bodhisattva Master-Saint, a Maha-Deva. He is centered within the hub or center of the Great-Soul Medallion of the White Line of Masters. The Venerable One correlates to Bartholomew, the disciple of Jesus. The Venerable One was Diogenes. When a disciple is aware of his need to unite with the Mediation of the Higher Worlds, his evolvement permits him to come under the discipline and instantaneous reproving of the Venerable One.

Brhaspati, one of the Great Immortals, works directly with Jesus through the Gunas. He works directly with our Father through prayer. Brhaspati works directly with Melchizedek through the tithe. Brhaspati works directly with Maha Chohan to restore and quicken new human-era cycles in religion. He is assisted by Joseph, the father of Jesus.

The Venerable One correlates to the Second Light-Stream (Devotion). With Joseph, the father of Jesus, he is in charge of all twelve Light-Streams. Presently, the Venerable One opens seven of the Light-Streams to one another.

VISHNU—The Father; the Conditioned or Personalized God.

VISHUDDHA—The Fifth Chakra centered in the throat.

WHITE BROTHERS—The White Brothers work with the Luminosity Angels of the dead so that man may in death make transition in an acceptable and natural process. This part of the Cosmos Family dwells in the total range of the Astral World. It is their work to aid certain of the dead who are confined in the subterranean corridors of the Astral World to rise into higher degrees of Light. Jesus spoke of the White Brothers when He said, *"Let the dead bury their dead." (St. Luke 9:60)* In the book of Acts, they are referred to as the "men . . . in white apparel." (Acts 1:10,11)

WORLD SOUL—The combined lower and higher subconscious impulses of all sentient and consciousness life of the earth united with the Love of God, under command of Him who is the Regent of all souls of the earth—Jesus.

YANG—Positive or masculine polarity correlating to Shiva.

YANTRA—A Yantra is a pattern or design to lead one to a spiritual idea, and thus produces a breakthrough in consciousness. It is usually experienced as a geometrical design. The yantra power signifies that one has opened the Bindu, or Mouth-of-God Center.

YIN—The feminine side of polarity; Shakti energy power.

INDEX

B

U

V

W

OTHER BOOKS BY ANN REE COLTON

THE THIRD MUSIC
A powerful book describing the energy-worlds of the Mind, the Soul and the Universe.

KUNDALINI WEST
Knowledge of the Kundalini and the Chakras for the Western initiate.

WATCH YOUR DREAMS
An invaluable and necessary book revealing the soul-codes in dreams and their symbols.

ETHICAL ESP
An important book defining the difference between lower and higher ESP.

THE JESUS STORY
A miracle book in timing to the need for miracles.

THE HUMAN SPIRIT
A scientific, spiritual, and healing book on the creation, purpose and destiny of man.

THE SOUL AND THE ETHIC
A profound book on the soul and on the ethical use of soul-power.

THE KING
From the personal, hieroglyphic journal of Ann Ree Colton.

DRAUGHTS OF REMEMBRANCE
An extraordinary book on the subject of reincarnation.

MEN IN WHITE APPAREL
A book of vital revelations about death and the life after death.

THE VENERABLE ONE
An initiatory book for those who love Nature and who would unveil Nature's secrets.

VISION FOR THE FUTURE
A prophetic book to comfort men in a perilous time.

THE LIVELY ORACLES
A prophetic book on world events.

ISLANDS OF LIGHT
A book of initiation with an underlying prophetic theme.

PRECEPTS FOR THE YOUNG
Appreciated by the adult . . . inspiring to the child . . . beneficial to the family.

BOOKS CO-AUTHORED
BY ANN REE COLTON
AND JONATHAN MURRO

PROPHET FOR THE ARCHANGELS
The Life Story of Ann Ree Colton.

THE PELICAN AND THE CHELA
The Teacher-Student relationship in the spiritual life.

GALAXY GATE I: THE HOLY UNIVERSE
A remarkable book of spiritual revelations about Man, the Solar System and the Cosmos.

GALAXY GATE II: THE ANGEL KINGDOM
A book filled with enlightening insights into the World of Holy Mediators between God and mankind.

OWE NO MAN
Scriptural Principles of Good Stewardship and Divine Providence.

THE ANOINTED
Sacred Keys to Healing, Exorcism and the Divine Marriage.

ANN REE COLTON FOUNDATION
Post Office Box 2057 Glendale, California 91209